Published by

World Scientific Publishing Co. Pte. Ltd.
5 Toh Tuck Link, Singapore 596224
USA office: 27 Warren Street, Suite 401-402, Hackensack, NJ 07601
UK office: 57 Shelton Street, Covent Garden, London WC2H 9HE

Library of Congress Cataloging-in-Publication Data
Vaisman, Izu., 1938–
Analytical geometry / Izu Vaisman.
p. cm. -- (Series on University Mathematics; vol. 8)
Includes bibliographical references (pp. 279–280) and index.
ISBN-13 978-981-02-3158-3 -- ISBN-10 981-02-3158-X
ISBN-13 978-981-256-857-1 (pbk) -- ISBN-10 981-256-857-3 (pbk)
1. Geometry, Analytic. I. Title. II. Series.
QA551.V24 1997
516.3--dc21 97-20206
CIP

British Library Cataloguing-in-Publication Data
A catalogue record for this book is available from the British Library.

Series On University Mathematics Vol. 8

ANALYTICAL GEOMETRY

Izu Vaisman
Department of Mathematics
University of Haifa
Israel

World Scientific
Singapore • New Jersey • London • Hong Kong

ANALYTICAL GEOMETRY

SERIES ON UNIVERSITY MATHEMATICS

ISSN: 1793-1193

From Tzuong-Tsieng Moh, a long-time expert in algebra, comes a new book for students to better understand linear algebra. Writing from an experienced standpoint, Moh touches on the many facets surrounding linear algebra, including but not limited to, echelon forms, matrix algebra, linear transformations, determinants, dual space, inner products, the Gram–Schmidt Theorem, Hilbert space, and more. It is ideal for both newcomers and seasoned readers who want to attain a deeper understanding on both the basics and advanced topics of linear algebra and its vast applications. The wide range of topics combined with the depth of each discussion make it essential to be on the shelf of every mathematical beginner and enthusiast.

Published

Vol. 1 Lectures on Differential Geometry
by S S Chern, W H Chen and K S Lam

Vol. 2 Lectures on Lie Groups
by W-Y Hsiang

Vol. 5 Algebra
by T-T Moh

Vol. 6 A Concise Introduction to Calculus
by W-Y Hsiang

Vol. 7 Number Theory with Applications
by W C Winnie Li

Vol. 8 Analytical Geometry
by Izu Vaisman

Vol. 9 Lectures on Lie Groups (Second Edition)
by W-Y Hsiang

Vol. 10 Linear Algebra and its Applications
by T-T Moh

Foreword

The more or less standard American (and other) curriculum for college and university students in mathematics and related fields (computer science, physics, engineering, etc.) includes courses on Basic Linear Algebra, and Calculus and Analytical Geometry in the first year. In particular, such courses provide the student with a first knowledge on vectors and coordinates, as well as a superficial acquaintance with conic sections and quadrics. Later on, those who develop an interest in Geometry may take second courses in this field and, mainly, such second courses will discuss axiomatic and non-Euclidean geometry. (Though false, sometimes the word geometry is not even mentioned in these courses, I refer to the analytical geometry part of the first year curriculum as a first course in geometry.)

In this way such classical topics as the geometric study and classification of conics and quadrics, and projective geometry are not covered. But, conics and quadrics, in spite of being old, are of great importance in both pure and applied mathematics. (After all, the trajectories of the planets are conics!) The classification of conics and quadrics yields an excellent example of a typical problem of geometry: the classification of geometric objects, and the study of geometric invariants. And projective geometry, not only is a beautiful mathematical subject, but also is widely used in computer graphics.

In this book I propose an alternative way of teaching either a first or a second course in geometry, which would develop analytical affine and Euclidean geometry, also including geometric transformations, and give an introduction to projective geometry. This course would enhance the geometric education of the students, and benefit those who are interested in applications, computer graphics in particular. Let me also add that such a course continues to be rather standard for the first year curriculum of many continental European countries, France, Russia, etc.

Furthermore, in this book we also intend to advocate a return to the geometric patterns, teach students to perceive geometry as a world in itself, and develop their geometric intuition. It is well known since Hilbert's work on the foundations of geometry at the end of the 19th century that linear algebra can be deduced from Euclidean geometry: the geometry can be used as the starting point in the construction of the real n-dimensional vector space (see, for instance, [14]). $\mathbf{R}^2$ and $\mathbf{R}^3$ are only models of plane and space.

It is true that foundations are a problem from this point of view. Foundation of geometry is not a beginner's subject, and instead of giving an incomplete axiomatic development, we will prefer to build on the undefined but basic geometric intuition. (This is done, usually, in first year calculus too when one says that a vector is a sequence of three numbers, and one draws an arrow, and also when one works with real numbers without a proper definition of these numbers.) We will take care to keep the latter to a minimal, harmless level. The gain in geometric thinking is more important than the temporary loss of rigour in foundations, which, however, should be openly explained to the students. In the book, we also indicate a possible solution to the foundations' problem by giving a logically rigorous definition of affine and Euclidean space which uses the algebraic structure of a linear space (Section 1.5), but we do not dwell much on this subject.

The book is written in such a way that it can be studied in either the first or the second year. The use of linear algebra is kept to a minimum which the teacher can provide by himself, if necessary. If taught in the second year, the students will already have had some linear algebra, calculus and the rudiments of analytical geometry. They will be familiar with orthogonal coordinates, vectors, lines, planes, and the simplest equations of conics and quadrics. They will also be more mature, mathematically, and used to the fact that mathematics consists of definitions, theorems and proofs, as well as with the logical structure of a proof.

Nevertheless, since the study of analytical geometry in the framework of the courses on linear algebra or calculus is rather cursory, and far from being uniform, we discuss the subject from the beginning. This allows for the establishment of the geometric methodology, while the previous exposure of the student to vectors and coordinates will provide a basis for understanding this new methodology.

The book starts with a chapter on vectors, and general affine coordinates. Then, we discuss equations of straight lines and planes, and use these equations for solving geometric problems. A teacher who feels that his/her students already know these subjects well may go through these chapters at a quick pace but, I would strongly recommend to spend some time on solving geometric problems. The same is true for parts of the chapter on conics and quadrics which consists mainly of new material, however. I also emphasize that our study of geometry is in space and not just in the plane. Many second courses in geometry, study only plane geometry, and the difference in what concerns the development of the geometric thinking and intuition of the students is tremendous. The chapter on conics and quadrics, which I call quadratic geometry, gives a complete, geometrically based solution to the classification problem, from the affine and the Euclidean point of view, including the notion of an invariant, and the basic invariants of conics and quadrics. The methodology of the classification includes some novelties.

The next chapter develops the subject of geometric transformations, and we tried to make it as simple as possible. We establish the most important properties of affine and orthogonal transformations, and prove that the orthogonal transformations are composed of symmetries.

In the last chapter, we give an introduction to projective geometry, again mainly in the three-dimensional case. We start with the enlarged affine space, then introduce projective coordinates by applying a general linear transformation to the homogeneous affine coordinates. We explain how to use projective coordinates in solving linear and quadratic geometric problems, including the projective classification of conics and quadrics, and give a cursory introduction to projective transformations.

I am convinced that the book provides a useful textbook for a second course in geometry for many categories of students: in mathematics, computer science, physics, engineering, etc., and I hope that it will contribute to a revival of the study of geometry in colleges and universities. The book may be used for individual study but, the help of a qualified instructor would make the study much easier. Indeed, while we are keeping at an elementary level, and the technical terms and notation are well explained, the text has a density which grows with the advancement, and asks for a corresponding understanding effort.

The classical character of the material included in this book, makes an extensive bibliography unnecessary. Needless to say, the book contains no original results. My main source for the theoretical part was [7]. The exercises and problems are also taken from older books, from [16] and [1], in particular, but from many other sources as well. At the end of the book, rather complete solutions of all these exercises and problems are given.

Let me also make a few more quotations: [15] for advanced analytical and projective geometry, [8], [9] for textbooks on projective geometry, [10] for an ample treatise of projective geometry, [13] for a classical treatise in English, [2] [3] for a modern, advanced treatise of geometry, [12] for an algebraically founded course of analytical geometry, [14] for a course on the foundations of geometry and [11] for applications to computer graphics.

We end the book with an Appendix concerning the utilization of the packages of Geometry and Projective Geometry of MAPLE, for the benefit of the students who would like to use a computer as an auxiliary of their study. But, please, do not neglect to use your own thinking in problem solving!

Contents

Chapter 1

The Algebra of Vectors

In this course, when we refer to Geometry, what we have in mind is the study of *points, lines, planes* and other *figures* (i.e., *sets of points*)[1] as they appear to us in the high school and college courses of Geometry. Figures that sit in a plane belong to the realm of *plane* or *two-dimensional geometry*, while those which have *thickness* belong to *solid* or *three-dimensional geometry* of *space*. It is also possible to refer to *one-dimensional geometry* as geometry along a straight line.

Furthermore, the geometry we think about here is called *Euclidean Geometry* (*Euclidean plane*, *Euclidean space*), because it was Euclid (3rd century BC) who first systematized its study in his famous book "Elements".

Analytical Geometry (Euclidean or others) is the study of geometry by algebraic means. It was discovered by René Descartes (1596-1650), a famous French mathematician and philosopher, who explained the method in his book *Géometrie* (1637).

Essentially, the method of Analytical Geometry consists in the construction of a *dictionary* which gives an algebraic translation of the geometric problems, and a geometric translation of the algebraic solution of these problems. What we intend to study is this dictionary, precisely.

The most important entry of our dictionary is *point* $\longleftrightarrow$ *coordinate numbers*, and this was in Descartes' book. However, in the 19th

[1] In the book, words in italics are words which either define new names or we intend to emphasize for one reason or another.

century, and for the needs of physics in particular, Hamilton (England) and Grassmann (Germany) discovered some intermediate objects called *vectors*, which are geometric objects, on one hand, and which may be computed with, on the other hand. The vectors are very good to represent physical notions such as *velocity, force*, etc., but this representation is beyond our scope.

Accordingly, we are going to study the *algebra of vectors* first and, then, we will use it to develop Analytical Geometry.

1.1 Free Vectors

We assume that the reader is familiar with elementary college geometry. Here are a few basic notions which we need in the study of vectors.

Geometric *points* may *move* along *straight lines* in one of the two possible *directions* (e.g., East-West or West-East). A straight line with a chosen direction is called an *axis*; the chosen direction will be called *positive*, and the opposite direction will be *negative*. A direction is a common feature of the whole family of straight lines which are *parallel* with a given line. For instance, the East-West direction of a map may be represented on anyone of the parallel lines which represent a constant geographic latitude.

Sometimes, it is necessary to refer to a couple of opposite directions (e.g., East-West and West-East) as to *one* mathematical object. This object is identifiable with a family of parallel straight lines. We will refer to it as an *unmarked direction.* Then, a usual direction will be called a *marked direction.* If no special mention is made, all our directions will be marked.

Geometric figures may be moved in plane or space in such a way that all the points of the figure move along parallel lines, into the same direction and by the same length. Such a motion is called a *parallel translation.*

The simplest geometric figures are the *straight line segments.* A *segment* AB is defined precisely by the pair of points (A, B) which are the *endpoints* of the segment. If a length unit is chosen, and whenever we speak of length in this book we assume that such a choice was made, every segment has a well-defined *length* $l(AB)$ which is a non-negative real number. An *ordered* pair of points (A, B) defines an *oriented segment.* (A usual segment is *nonoriented.*) This oriented segment will be denoted by $\vec{AB}$, where A is the *origin* and B is the *end* of the segment. The orientation of the segment yields a well-defined direction called the *direction of the segment.* If the oriented segment lies on an axis, we say that the segment is *positive* or *negative* according to the fact that the direction of the segment coincides with the positive or negative direction of the axis. We will define the notion of *algebraic length* of an oriented segment of an axis by the number $a(\vec{AB}) = \pm l(AB)$, where we take $+$ for the positive segments and $-$ for the negative segments.

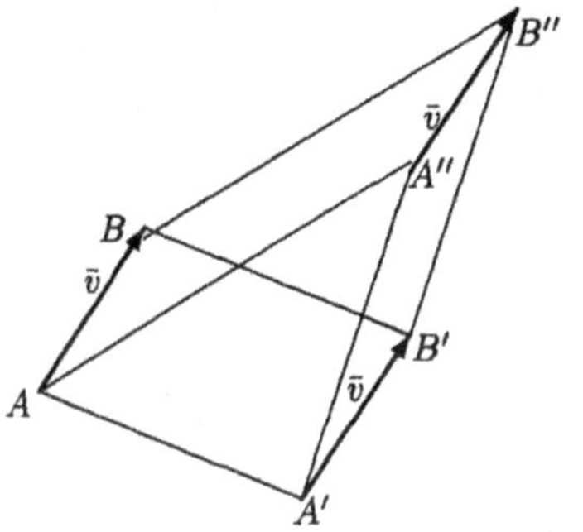

Fig. 1.1.1

With all these terminologies put in place, we can now define:

1.1.1 Definition. *A free vector $\bar{v}$ is a class of oriented segments, $\vec{AB}$, etc. defined up to parallel translation. Anyone of the segments of the class of oriented segments obtained by translating $\vec{AB}$ parallel is called a representative segment of $\bar{v}$. In other words, $\vec{AB}$ represents $\bar{v}$ at the origin A and with endpoint B, and we will also write $\bar{v} = \vec{AB}$.*

Thus, we must understand that, if $\vec{A'B'}$ is obtained from $\vec{AB}$ by a parallel translation, the vectors defined by these two segments are regarded as being equal. (See Fig. 1.1.1.) Obviously, this situation occurs iff (this is not a spelling error; mathematicians use "iff" instead of the words "if and only if") $\vec{AB}$ and $\vec{A'B'}$ have equal *length* and *direction.* It is important to notice that any two oriented segments $\vec{A'B'}$, $\vec{A''B''}$ of the class which defines a vector $\bar{v}$ are obtainable one from the other by a parallel translation. Indeed, if $\bar{v}$ is first represented by $\vec{AB}$, then $\vec{A'B'}$ and $\vec{A''B''}$ have equal length and direction namely, the length and direction of $\vec{AB}$ (see Fig. 1.1.1 once more, hence $\vec{A''B''}$ may be obtained by a parallel translation of $\vec{A'B'}$).

The common length and direction of all the representative segments of a vector are called the *length* and *direction* of the vector itself. The length of the vector $\bar{v}$ will be denoted by $|\bar{v}|$. A vector of length 0 will be called a *zero vector*, and denoted by $\bar{0}$. Clearly the corresponding class of oriented segments have the endpoint and origin at the same point. Hence, we may either say that we have a zero vector in every direction or, more conveniently, say that there is one zero vector whose direction is arbitrary. (The direction of a nonzero vector is always well-defined.) A vector of length 1 is called a *unit vector.* For any vector $\bar{v}$, there exists a well-defined vector which has the same length as $\bar{v}$ but opposite direction. It will be called the *opposite vector*, and denoted by $-\bar{v}$. If $\bar{v}$ is represented by $\vec{AB}$, $-\bar{v}$ may be represented by $\vec{BA}$.

1.1.2 Proposition. *i) There exists one and only one vector $\bar{v}$ associated to a given ordered pair of points (A, B). ii) For any vector $\bar{v}$, and any given point A, there exists one and only one point B such that $\vec{AB}$ represents the vector $\bar{v}$.*

Proof. i) We just take $\bar{v}$ to be defined by $\vec{AB}$. ii) B is the point situated on a parallel line to $\bar{v}$, in the direction of $\bar{v}$, and at a distance $|\bar{v}|$ from A. Q.E.D.

Let us emphasize once again that all our vectors will be free vectors. The fact that a vector is characterized by length and direction makes vectors so useful in physics, and, there, one also uses other kinds of vectors sometime, e.g. with a fixed origin or sliding along a straight line only.

Because we work with free vectors, only the relation of parallelism is significant for their position. Thus, two (or more) vectors will be called *colinear vectors* if they are parallel, and three (or an arbitrary family of) vectors will be called *coplanar vectors* if they are all parallel to a plane. In particular, if d is a given straight line, we will denote by $\mathcal{V}_d$ the set of all vectors that are parallel (or *belong*) to d, and if α is a given plane, we will denote by $\mathcal{V}_\alpha$ the set of all vectors which are parallel (*belong*) to α. The set of all free vectors of space will be denoted by $\mathcal{V}$.

Finally, notice again our utilization of the bar and the arrow in the notation of vectors, and do not forget it as you might be confused between vectors and numbers. It is the numbers which will be denoted just by simple letters.

1.2 Linear Operations with Vectors

The importance of vectors in geometry is due to the possibility of doing algebraic operations with vectors, and this possibility follows from the fact that our vectors are free. First, we define

1.2.1 Definition. *Let $\bar{u}$, $\bar{v}$ be two vectors, and represent them as $\bar{u} = \vec{AB}$, $\bar{v} = \vec{BC}$, where A is taken arbitrarily. Then, the vector defined by $\vec{AC}$ is called the sum of $\bar{u}$, $\bar{v}$, and it is denoted by $\bar{u} + \bar{v}$.*

The sum $\bar{u} + \bar{v}$ does not depend on the choice of A as the reader can easily see from Fig. 1.2.1. If A is chosen, B and C are well-defined by Proposition 1.1.2. If the vectors $\bar{u}$, $\bar{v}$ are not parallel, $\bar{u} + \bar{v}$ can also be obtained by the so-called *rule of the parallelogram* (as opposite to Definition 1.2.1, which is the *triangle rule*): represent $\bar{u} = \vec{AB}$ and $\bar{v} = \vec{AC}$; then $\bar{u} + \bar{v} = \vec{AD}$, where D is the opposite vertex of A in the parallelogram $ABDC$ (Fig. 1.2.2).

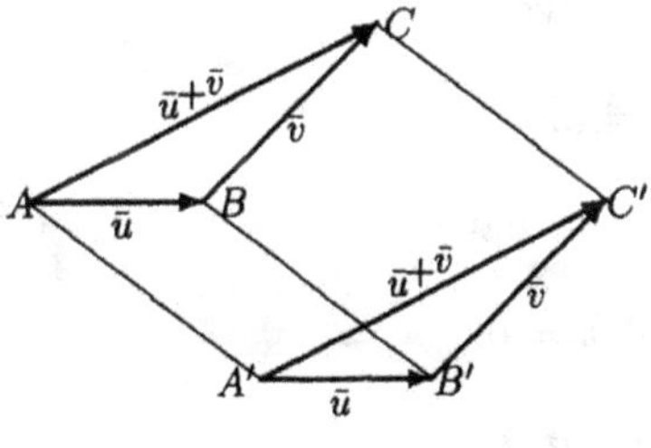

Fig. 1.2.1

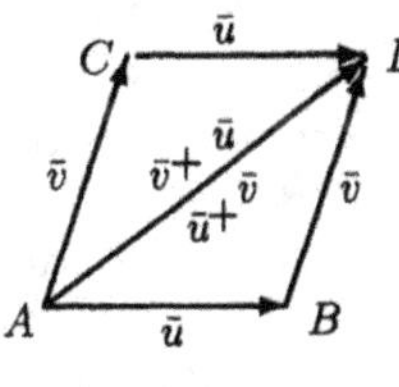

Fig. 1.2.2

Adding vectors is something different from adding their lengths. Indeed, we have

1.2.2 Proposition. *For any two vectors $\bar{u}$, $\bar{v}$, the following relation*

holds:

$$|\bar{u}+\bar{v}| \leq |\bar{u}|+|\bar{v}|. \tag{1.2.1}$$

In (1.2.1), equality holds iff $\bar{u}$ and $\bar{v}$ are parallel and have the same direction.

Proof. Indeed, in the triangle ΔABC of Fig. 1.2.1 the lengths of the edges must satisfy $AB+BC \leq AC$, with equality holding iff the triangle degenerates to the line segment AC with B lying on this segment. Q.E.D.

1.2.3 Remark. *From another known property of the same triangle ΔABC, we get*

$$||\bar{u}|-|\bar{v}|| \leq |\bar{u}+\bar{v}|, \tag{1.2.2}$$

with equality holding iff $\bar{u}$ and $\bar{v}$ are parallel and have opposite directions.

Now, we present the basic computational properties of the sum of vectors:

1.2.4 Proposition. a) *Associativity: for any three vectors we have*

$$(\bar{u}+\bar{v})+\bar{w}=\bar{u}+(\bar{v}+\bar{w}). \tag{1.2.3}$$

b) *Existence of zero: for any vector $\bar{u}$ we have*

$$\bar{u}+\bar{0}=\bar{0}+\bar{u}=\bar{u}. \tag{1.2.4}$$

c) *Existence of opposite: for any vector $\bar{u}$ we have*

$$\bar{u}+(-\bar{u})=(-\bar{u})+\bar{u}=\bar{0}. \tag{1.2.5}$$

d) *Commutativity: for any two vectors we have*

$$\bar{u}+\bar{v}=\bar{v}+\bar{u}. \tag{1.2.6}$$

Proof. The proof of associativity is shown in Fig. 1.2.3. The reader can easily prove alone the properties *b*) and *c*) using the definition of the sum. Finally, if $\bar{u}$ and $\bar{v}$ are not parallel, commutativity is proven in Fig. 1.2.2, and, if they are parallel, (1.2.6) follows by comparing the length and direction of the two sides of the formula. Q.E.D.

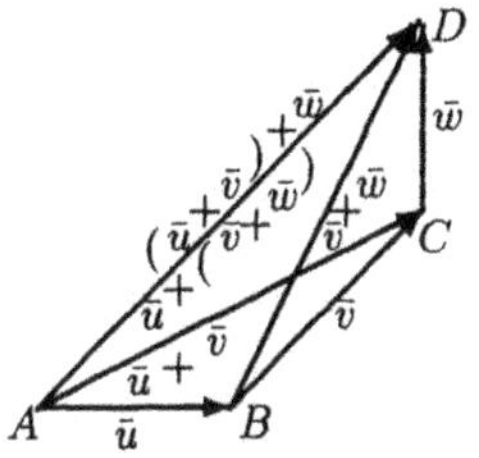

Fig. 1.2.3

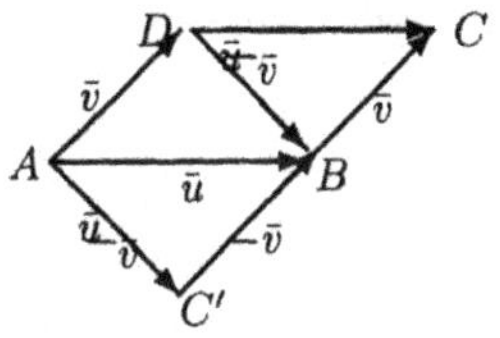

Fig. 1.2.4

1.2.5 Remark. *We recall that an algebraic (i.e., computational) system closed with respect to an operation which satisfies the properties a), b), c) above is called a group. If property d) also holds, the group is said to be commutative or Abelian (from the name of the Norwegian 19th century mathematician Niels Abel).*

The importance of associativity lies in the fact that it allows us to define the sum of any finite number of vectors. The result will be denoted by $\bar{u}_1 + \bar{u}_2 + \cdots + \bar{u}_n$, and it will be defined by inserting parentheses in this expression so as to add only two vectors each time. Associativity ensures that the result is independent of the choice of places for the parentheses.

1.2.6 Example. *If $A_1A_2 \cdots A_{n-1}A_n$ is a polygonal line, $\overrightarrow{A_1A_2} + \overrightarrow{A_2A_3} + \cdots + \overrightarrow{A_{n-1}A_n} = \overrightarrow{A_1A_n}$. In particular, if the polygonal line is closed, i.e. $A_n = A_1$, the sum of the vectors along its edges, in a fixed (clockwise or counterclockwise) sense equals $\bar{0}$.*

Next, the existence of the opposite vector allows us to define the operation of *difference of two vectors* in the following way:

$$\bar{u} - \bar{v} = \bar{u} + (-\bar{v}). \tag{1.2.7}$$

Figure 1.2.4 shows two constructions of $\bar{u} - \bar{v}$, one by means of the triangle $\Delta ABC'$, and the other by means of the diagonal DB of the parallelogram $ABCD$. Of course, the second construction is usable only if the vectors are not parallel.

It is important to notice that the first way of writing the difference in Fig. 1.2.4 yields

$$\vec{AC'} = \vec{AB} - \vec{BC},$$

and, by using the properties in Proposition 1.2.4 conveniently (exercise for the reader!), this result is seen to be equivalent to

$$\vec{BC'} = \vec{AC'} - \vec{AB}.$$

This result is so important that we will formulate it as

1.2.7 Proposition. *For any three points A, B, C in space we have the vectorial relation*

$$\vec{BC} = \vec{AC} - \vec{AB}. \tag{1.2.8}$$

For another way of writing the same relation, which will be very important later on, we first define

1.2.8 Definition. *Let us choose an arbitrary point O as a general origin. Then, the vector $\bar{r}_M := \vec{OM}$ is called the radius vector of the point M.*

(In this book, the symbol := means *equal by definition.*)

Now, Proposition 1.2.7 may be expressed as

1.2.9 Proposition. *If O is a fixed origin, for any two points M, N one has*

$$\vec{MN} = \bar{r}_N - \bar{r}_M. \tag{1.2.9}$$

The next basic operation which we want to study is multiplication of a vector by a real number. In the context of vector operations, the real numbers are also called *scalars.* The set of all the real numbers is denoted by **R**.

1.2.10 Definition. *For any scalar λ and any vector $\bar{v}$, we denote either by $\lambda\bar{v}$ or $\bar{v}\lambda$ the vector whose length is $|\lambda\bar{v}| = |\lambda||\bar{v}|$ ($|\lambda|$ is the absolute value of λ), and whose direction is the same as the direction of $\bar{v}$ if $\lambda \geq 0$, and opposite to the direction of $\bar{v}$ if $\lambda < 0$. The vector $\lambda\bar{v}$ is the product of λ and $\bar{v}$.*

Before going on, we recall a few convenient mathematical symbols. Namely, $\forall$ means "for all", $\exists$ means "exists", $\in$ means "belongs to", and $\cap$ means "intersection".

The basic properties of the operation defined in Definition 1.2.10 are contained in

1.2.11 Proposition. *$\forall \bar{v}, \bar{w} \in \mathcal{V}$ and $\forall \lambda, \mu \in \mathbf{R}$, the product between a scalar and a vector satisfies:*
a) $1\bar{v} = \bar{v}$,
b) $\lambda(\mu\bar{v}) = (\lambda\mu)\bar{v}$,
c) $(\lambda + \mu)\bar{v} = \lambda\bar{v} + \mu\bar{v}$,
d) $\lambda(\bar{v} + \bar{w}) = \lambda\bar{v} + \lambda\bar{w}$. (The properties c) and d) are distributivity properties.)

Proof. a) is trivial, and b), c) are to be checked for all possible signs of λ and μ; in all the cases, it follows easily from Definition 1.2.10 that the vectors of the two sides of b) and c) have the same length and the same direction. Finally, d) follows from the similarity of the triangles $\triangle ABC$ and $\triangle AB'C'$ in Fig. 1.2.5. Q.E.D.

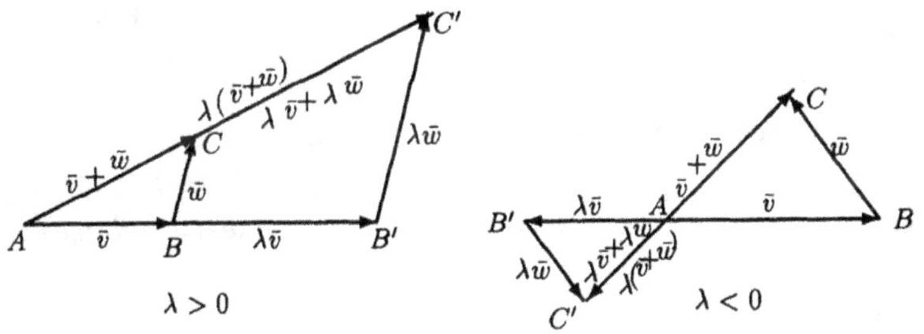

Fig. 1.2.5

1.2.12 Remark. *We recall that an algebraic (i.e., computational) system consisting of a set of elements called (abstract) vectors, which is an Abelian group for some abstract addition operation, and where the vectors may be multiplied by real (scalar) numbers, by an operation which yields a new vector, and satisfies the properties a), b), c), d) of Proposition 1.2.11 is called a (real) linear space or a (real) vector space. For instance the sets of vectors $\mathcal{V}_d$, $\mathcal{V}_\alpha$, $\mathcal{V}$ defined earlier in this section are linear spaces.*

We also recall the following very important definition

1.2.13 Definition. *i) An expression of the type*

$$\lambda_1\bar{v}_1 + \lambda_2\bar{v}_2 + \cdots + \lambda_h\bar{v}_h \tag{1.2.10}$$

is called a linear combination of the vectors $\bar{v}_1, \bar{v}_2, \cdots, \bar{v}_h$.
ii) If the result of the linear combination (1.2.10) cannot be $\bar{0}$ *unless* $\lambda_1, \lambda_2, \cdots \lambda_h$ *are all equal to 0, the vectors* $\bar{v}_1, \bar{v}_2, \cdots \bar{v}_h$ *are said to be linearly independent. If, on the contrary, there exists a set of coefficients* λ*, not all of them 0, such that (1.2.10) vanishes, the vectors* $\bar{v}_1, \bar{v}_2, \cdots, \bar{v}_h$ *are linearly dependent.*

The following facts follow straightforwardly from Definition 1.2.13:
a) If one of the vectors $\bar{v}_i$, $i = 1, \ldots, h$ is $\bar{0}$, the vectors which enter in (1.2.10) are linearly dependent. (If $\bar{v}_i = \bar{0}$, we may take $\lambda_i \neq 0$, and $\lambda_j = 0$ for $j \neq i$, and the combination (1.2.10) vanishes.)
b) If the vectors $\bar{v}_1, \bar{v}_2, \cdots, \bar{v}_h$ are linearly dependent, and if we add to their system any other vectors $\bar{v}_{h+1}, \cdots, \bar{v}_k$, the vectors $\bar{v}_1, \bar{v}_2, \cdots, \bar{v}_k$ are again linearly dependent. (If we add the new vectors multiplied by 0 to (1.2.10), and if there were nonzero coefficients in (1.2.10) such that the result was zero, this situation remains unchanged.) c) A subsystem of a system of linearly independent vectors consists itself of linearly independent vectors. (Otherwise, b) above would be contradicted.)

1.2.14 Proposition. *The vectors* $\bar{v}_1, \bar{v}_2, \cdots, \bar{v}_h$ *are linearly independent iff one of these vectors can be expressed as a linear combination of the others.*

Proof. If

$$\lambda_1\bar{v}_1 + \lambda_2\bar{v}_2 + \cdots + \lambda_h\bar{v}_h = \bar{0}$$

and $\lambda_1 \neq 0$, for instance, we get

$$\bar{v}_1 = \mu_2\bar{v}_2 + \mu_3\bar{v}_3 + \cdots + \mu_h\bar{v}_h, \tag{*}$$

where $\mu_i = -(\lambda_i/\lambda_1)$, $i = 2, 3, \cdots, h$.

Conversely, it is clear that (*) is equivalent to a relation of linear dependence since the coefficient of $\bar{v}_1$ is nonzero. Q.E.D.

1.2.15 Proposition. *Two vectors are linearly dependent iff they are colinear or, equivalently, iff they are proportional.*

Proof. Proposition 1.2.14 shows that $\bar{u}$ and $\bar{v}$, are linearly dependent iff $\bar{u} = \lambda\bar{v}$, say, i.e., the two vectors are proportional. And, proportionality is equivalent to collinearity because of Definition 1.2.10. Q.E.D.

1.2.16 Proposition. *Three vectors are linearly dependent iff they are coplanar vectors (i.e., they belong to a plane).*

Proof. $\bar{u}$, $\bar{v}$, $\bar{w}$ are linearly dependent iff, say, $\bar{u} = \lambda\bar{v} + \mu\bar{w}$, and the geometric definition of the vector operations shows that $\bar{u}$ belongs to (i.e., is parallel with) the plane determined by $\bar{v}$, $\bar{w}$.

Conversely, if two of the given vectors are collinear, they are already linearly dependent hence the same holds for all three vectors. Let us look at three coplanar vectors $\bar{u}$, $\bar{v}$, $\bar{w}$, no two of which are on a same line. Then, we can construct in an obvious way the parallelogram $ABCD$ of Fig. 1.2.6 which yields the relation of linear dependence $\bar{w} = \lambda\bar{u} + \mu\bar{v}$. Q.E.D.

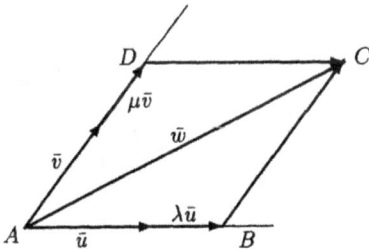

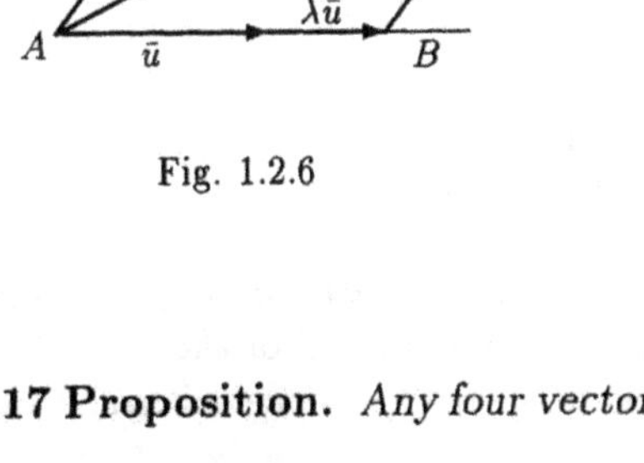

Fig. 1.2.6

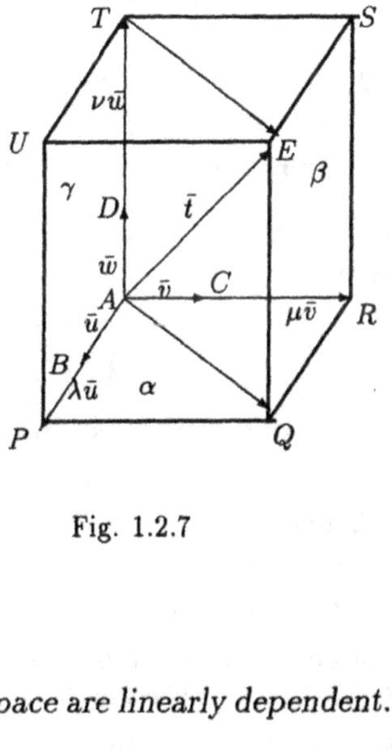

Fig. 1.2.7

1.2.17 Proposition. *Any four vectors of space are linearly dependent.*

Proof. If three of the vectors are already dependent, all four are. Let us look at the vectors $\bar{u} = \vec{AB}$, $\bar{v} = \vec{AC}$, $\bar{w} = \vec{AD}$, $\bar{t} = \vec{AE}$, no three of which are in the same plane. Then the planes α, β, γ defined by the triples ABC, ACD, ADB, respectively, and the planes through E and parallel to α, β, γ define the parallelipiped $APQRSEUT$ of Fig. 1.2.7, and we see that

$$\bar{t} = \vec{AE} = \vec{AQ} + \vec{AT} = \vec{AP} + \vec{AR} + \vec{AT} = \lambda\bar{u} + \mu\bar{v} + \nu\bar{w}.$$

Q.E.D.

We end this section with two solved problems and unsolved problems for the reader. In order to formulate the first problem, we define the following important notion:

1.2.18 Definition. *If A, B, C are three points of an axis, the real number*

$$\kappa = (A, B; C) := a(\vec{AC})/a(\vec{CB}) \tag{1.2.11}$$

is called the simple ratio of the triple A, B, C.

1.2.19 Problem. *With respect to an arbitrary origin, compute the radius vector $\bar{r}_C$ of the point C of (1.2.11) by means of the radius vectors $\bar{r}_A, \bar{r}_B$ of A and B and of the simple ratio κ.*

Solution. Definition (1.2.11) is equivalent to $\vec{AC} = \kappa\vec{CB}$. Hence, we have

$$\bar{r}_C - \bar{r}_A = \kappa(\bar{r}_B - \bar{r}_C), \tag{1.2.12}$$

and we obtain the required result

$$\bar{r}_C = \frac{\bar{r}_A + \kappa\bar{r}_B}{1+\kappa}. \tag{1.2.13}$$

1.2.20 Remark. *Notice that $\kappa = 1$ characterizes the midpoint of AB. Notice also that, if A and B are fixed, the points C of the straight line d defined by AB are in a one-to-one correspondence with the values $\kappa \neq -1$ of the simple ratio. For $C = B$, $\kappa = \infty$.*

1.2.21 Problem. *Prove the concurrence of the three medians of an arbitrary triangle.*

Solution. We look at the triangle ΔABC of Fig. 1.2.8, where A', B', C'

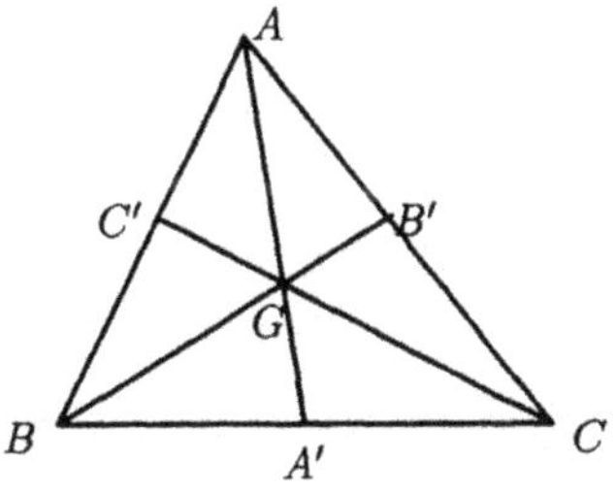

Fig. 1.2.8

are the midpoints of BC, CA, AB, respectively. Then, if we use the point A as an origin, and on the basis of formula (1.2.13), we get

$$\vec{AA'} = \frac{1}{2}(\vec{AB} + \vec{AC}), \quad \vec{AB'} = \frac{1}{2}\vec{AC}, \quad \vec{BB'} = \frac{1}{2}\vec{AC} - \vec{AB}.$$

Now, let $G = AA' \cap BB'$ denote the intersection point of the medians AA' and BB'. Then, $\vec{AG} = \lambda\vec{AA'}$ and $\vec{BG} = \mu\vec{BB'}$ for some scalars λ, μ. Moreover, we have $\vec{BG} = \vec{AG} - \vec{AB}$, i.e.

$$\mu(\frac{1}{2}\vec{AC} - \vec{AB}) = \frac{\lambda}{2}(\vec{AB} + \vec{AC}) - \vec{AB}.$$

The latest relation amounts to

$$(\frac{\lambda}{2} + \mu - 1)\vec{AB} - \frac{\lambda - \mu}{2}\vec{AC} = \bar{0},$$

and since $\vec{AB}, \vec{AC}$ are linearly independent, the coefficients must vanish and we obtain $\lambda = \mu = 2/3$. In particular, we have the simple ratio $(A, A'; G) = 2/3$.

Similarly, if $G' = AA' \cap CC'$, we will obtain $(A, A'; G') = 2/3$. By the second part of Remark 1.2.20, we obtain $G = G'$.

The point G of this problem is the *center of gravity (centroid)* of the triangle ΔABC. Using the value $(A, A'; G) = 2/3$, one obtains the following value of the radius vector of G with respect to any origin

$$\bar{r}_G = \frac{1}{3}(\bar{r}_A + \bar{r}_B + \bar{r}_C). \tag{1.2.14}$$

(Reader, please check!)

Exercises and Problems

1.2.1. Let $\bar{a}$, $\bar{b}$, $\bar{c}$ be linearly independent vectors. i) What can be said about the linear dependence of the vectors

$$\bar{l} = 2\bar{b} - \bar{c} - \bar{a}, \quad \bar{m} = 2\bar{a} - \bar{b} - \bar{c}, \quad \bar{n} = 2\bar{c} - \bar{a} - \bar{b}\ ?$$

ii) Write down the decomposition of $\bar{s} = \bar{a}+\bar{b}+\bar{c}$ as a linear combination of the vectors

$$\bar{l}' = \bar{a} + \bar{b} - 2\bar{c}, \quad \bar{m}' = \bar{a} - \bar{b}, \quad \bar{n}' = 2\bar{b} + 3\bar{c},$$

if it exists.

1.2.2. Consider the quadrilateral $ABCD$, and let E, N, F, M be the midpoints of the edges AB, BC, CD, DA respectively. Prove that

$$\vec{EF} = \frac{1}{2}(\vec{AD} + \vec{BC}), \qquad \vec{AC} = \vec{MN} + \vec{EF}.$$

1.2.3. For any given triangle, prove that there exists a triangle whose edges have the same lengths and unmarked directions as the medians of the initial triangle. If this construction is repeated twice, the second triangle is similar to the initial one with similarity ratio 3/4.

1.2.4. In any tetrahedron (i.e., a triangular pyramid), prove that the segments which join a vertex to the centroid of the opposite face, and the segments which join the midpoints of opposite edges have all one common point, called the *center of gravity (centroid)* of the tetrahedron.

1.2.5. Let $SABC$ be a tetrahedron, and let A', B', C' be arbitrary points of the edges SA, SB, SC, respectively. Let M be the intersection point of the planes $A'BC$, $B'CA$, $C'AB$, and N be the intersection point of the planes $AB'C'$, $BC'A'$, $CA'B'$. Prove that the points M, N, S are collinear points.

1.3 Coordinates of Vectors and Points

In this section, we use the results on vectors in order to make the first basic translation between geometry and algebra, namely we shall establish one-to-one correspondences between vectors and some numbers, called *coordinates*, as well as between points and *coordinates*. We treat this problem on a straight line d, in a plane α, and in space.

1.3.1 Definition. *With the notation of Remark 1.2.12, if $\mathcal{V}_d$ is the linear space of all the vectors of a straight line d, a* **basis of** $\mathcal{V}_d$ *(or a* **vector basis of** *d) is a nonzero vector of this space. If $\mathcal{V}_\alpha$ is the linear space of all the vectors of a plane α, a* **basis of** $\mathcal{V}_\alpha$ *(or a* **vector basis of** α*) is a pair of linearly independent (i.e., noncollinear) vectors of this space. If $\mathcal{V}$ is the linear space of all vectors of space, a* **basis of** $\mathcal{V}$ *(or a* **vector basis of space***) is a triple of linearly independent (i.e., noncoplanar) vectors of $\mathcal{V}$.*

The name *basis* is justified by

1.3.2 Proposition. *If $\bar{e}$ is a basis of $\mathcal{V}_d$, any vector $\bar{v} \in \mathcal{V}_d$ may be written in a unique way as*

$$\bar{v} = x\bar{e}, \qquad (x \in \mathbf{R}). \tag{1.3.1}$$

If $\bar{e}_1, \bar{e}_2$ is a basis of $\mathcal{V}_\alpha$, any vector $\bar{v} \in \mathcal{V}_\alpha$ may be written in a unique way as

$$\bar{v} = x_1\bar{e}_1 + x_2\bar{e}_2, \quad (x_1, x_2 \in \mathbf{R}). \tag{1.3.2}$$

If $\bar{e}_1, \bar{e}_2, \bar{e}_3$ is a basis of $\mathcal{V}$, any vector $\bar{v} \in \mathcal{V}$ may be written in a unique way as

$$\bar{v} = x_1\bar{e}_1 + x_2\bar{e}_2 + x_3\bar{e}_3, \quad (x_1, x_2, x_3 \in \mathbf{R}). \tag{1.3.3}$$

Proof. The existence of the decomposition (1.3.1), (1.3.2), (1.3.3) was proven in Propositions 1.2.15, 1.2.16, 1.2.17, respectively. The uniqueness is a straightforward consequence of the linear independence of the vectors of a basis. (Reader, please check!) Q.E.D.

1.3.3 Definition. *The coefficients x of the formulas (1.3.1), (1.3.2), (1.3.3) are called* **coordinates** *of the vector $\bar{v}$ with respect to the given basis. In particular, x of (1.3.1) and x_1 of the other two formulas is the* **abscissa**, *x_2 is the* **ordinate**, *and x_3 is the* **height**.

1.3.4 Remark. *We recall that, in any linear (vector) space, a basis is a maximal system of linearly independent vectors. Then, any vector is a unique linear combination of vectors of the basis, and the coefficients of this combination are the coordinates of the vector with respect to that basis. If finite, the number of vectors in a basis (which can be proven to be independent of the choice of the basis) is called the dimension or number of dimensions of the corresponding vector space. Thus $\mathcal{V}_d$ is one-dimensional, $\mathcal{V}_\alpha$ is two-dimensional and $\mathcal{V}$ is three-dimensional (it has three dimensions).*

1.3.5 Definition. *Generally, the bases defined by Definition 1.3.1 are called affine bases. But, if all the vectors of a basis are of length 1, and pairwise orthogonal (this is just another word for perpendicular), the basis will be called an orthonormal basis.*

We also mention that, sometimes we will prefer to write x instead of x_1, y instead of x_2, and z instead of x_3, and in the case of an orthonormal basis mostly, $\bar{i}$ instead of $\bar{e}$ and of $\bar{e}_1$, $\bar{j}$ instead of $\bar{e}_2$, and $\bar{k}$ instead of $\bar{e}_3$. If both notations appear in the same framework, the corresponding numbers and vectors are assumed to be equal.

1.3.6 Definition. *A configuration which consists of a point O chosen as origin and a basis of vectors is called a frame or a coordinate system on the line d, the plane α or in space, respectively. (Of course, in the case of a line or a plane, the origin is assumed to belong to that line or plane.) If the basis is affine, the frame is also affine, and if the basis is orthonormal, the frame is either called orthonormal or orthogonal. For any point M (of d, α or space, respectively), the coordinates of the radius vector $\bar{r}_M = \vec{OM}$ are called the coordinates of M with respect to the given coordinate system (frame).*

The coordinates of points, as defined above, may either be *affine* or *orthogonal*, according to the frame, and, in any case, they are also called *Cartesian coordinates*, in honor of Descartes, the inventor of Analytical Geometry, whose name in Latin was Cartesius.

The custom is to look at a frame as the figure which consists of the axes through the origin in the directions defined by the basic vectors, and, in space, also by the planes defined by each pair of axes. Accordingly, we speak of *coordinate axes* and *coordinate planes*, as shown in

Figs. 1.3.1 and 1.3.2 for plane and space, respectively. The case of a straight line is, say, the case of the x-axis in Fig. 1.3.1.

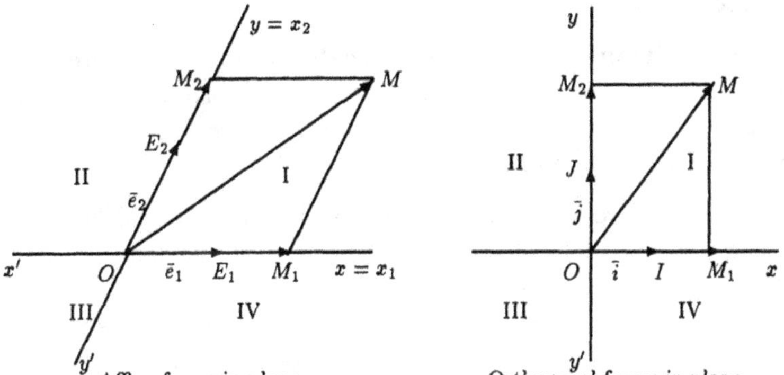

Affine frame in plane

Orthogonal frame in plane
$(|\bar{i}| = |\bar{j}| = 1,\ \bar{j} \perp \bar{i})$

Fig. 1.3.1

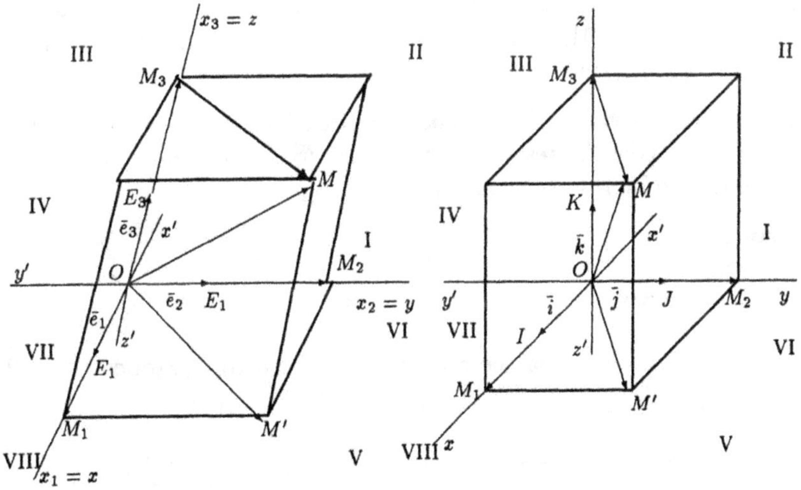

Affine frame in space

Orthogonal frame in space
$(|\bar{i}| = |\bar{j}| = |\bar{k}| = 1,$
$\bar{i} \perp \bar{j},\ \bar{j} \perp \bar{k},\ \bar{k} \perp \bar{i})$

Fig. 1.3.2

Furthermore, Figs. 1.3.1 and 1.3.2 also show us how to decompose the radius vector $\vec{OM}$ with respect to the corresponding bases by using parallels to the axes, in the affine case, and perpendiculars to the axes, in the orthogonal case. Moreover, the following interpretations of the coordinates hold

1.3.7 Proposition. *The affine coordinates of a point M are simple ratios: $x_M = x_{M_1} = -(M_1, E_1; O)$, and similar formulas for y_M and z_M. The orthogonal coordinates of a point M are algebraic lengths: $x_M = x_{M_1} = a(\vec{OM_1})$, and similar formulas for y_M and z_M.*

Proof. In this Proposition, the notation is that of Figs. 1.3.1 and 1.3.2, and the result follows straightforwardly from the definition of the simple ratio and the algebraic length. Q.E.D.

We will notice that the coordinate axes divide the plane into four *quadrants* numbered from I to IV, counterclockwise, starting from the one where the two coordinates are positive (see Fig. 1.3.1). The coordinate planes divide space into eight *octants*, four "upper" (with respect to the z-axis) octants numbered I to IV, counterclockwise, starting from the one where all the coordinates are positive, and four "lower" octants numbered V to VIII, counterclockwise (when looking from the upper z-axis), starting under octant I (see Fig. 1.3.2).

1.3.8 Proposition. *i) The coordinates of a sum of vectors are the sums of the coordinates of these vectors. ii) The coordinates of $\lambda\bar{v}$ are λ multiplied by the coordinates of $\bar{v}$.*

Proof. If

$$\bar{v} = v_1\bar{e}_1 + v_2\bar{e}_2 + v_3\bar{e}_3, \quad \bar{w} = w_1\bar{e}_1 + w_2\bar{e}_2 + w_3\bar{e}_3,$$

the known properties of the operations with vectors (Proposition 1.2.4 and Proposition 1.2.11) give

$$\bar{v} + \bar{w} = (v_1 + w_1)\bar{e}_1 + (v_2 + w_2)\bar{e}_2 + (v_3 + w_3)\bar{e}_3,$$

and

$$\lambda\bar{v} = (\lambda_1 v_1)\bar{e}_1 + (\lambda_2 v_2)\bar{e}_2 + (\lambda_3 v_3)\bar{e}_3.$$

Q.E.D.

1.3.9 Proposition. *Prove that for any two points A, B of an axis, where we have a unit basis and the coordinates are denoted by x, the following formula holds*

$$a(\vec{AB}) = x(B) - x(A).$$

Proof. As usual, $a(\vec{AB})$ is the algebraic length of the oriented segment. Since $\vec{AB} = \bar{r}_B - \bar{r}_A$, Proposition 1.3.8 tells us that $x(B) - x(A)$ is the coordinate of the vector $\vec{AB}$, which is exactly what we wanted to prove. Q.E.D.

1.3.10 Example. *If the simple ratio $(A, B; C) = \kappa$, then the space coordinates of C are given by*

$$x_C = \frac{x_A + \kappa x_B}{1+\kappa}, \quad y_C = \frac{y_A + \kappa y_B}{1+\kappa}, \quad z_C = \frac{z_A + \kappa z_B}{1+\kappa}.$$

Indeed, this follows from formula (1.2.13) and Proposition 1.3.8. In plane, we have no z, and on a line we have only x. If $\kappa = 1$, we get the midpoint of AB.

From the given definitions it is clear that if a coordinate system is chosen, every vector and every point have uniquely defined coordinates, and, conversely, given coordinates define a unique vector and a unique point, respectively.

But, it is just as clear that the same vector or point will have different coordinates with respect to different coordinate systems. Since we will use coordinates in order to solve geometric problems, the formulation of which does not depend on the choice of a coordinate system (this is the most important use of coordinates), it is very important to establish formulas for the *coordinate changes or coordinate transformations* imposed by a change of the coordinate system. Such formulas will then be used in checking the *invariance* of the geometric results established via coordinates.

Let $\{O; \bar{e}_i\}$ and $\{O'; \bar{e}'_i\}$ be two frames. We discuss the cases of a straight line, a plane, and space together. The only difference will be that indices like i are as follows: $i = 1$ for a line; $i = 1, 2$ for a plane, $i = 1, 2, 3$ for space. We do not indicate these values in the following formulas. The reader will have to do this in accordance with the relevant situation.

From Proposition 1.3.2, it follows that the two bases considered above are related by some concrete formulas (corresponding to the geometric definition of the indicated vectors) of the following type

$$\bar{e}'_i = \sum_j a_{ij}\bar{e}_j, \qquad \bar{e}_i = \sum_j b_{ij}\bar{e}'_j. \tag{1.3.4}$$

We recall that the symbol $\sum_j$ means the summation of the corresponding expression for all the necessary values of the *summation index* j. (If there are several summation indices, we perform summation following every one of these indices.) Moreover, if a summation index j is replaced by a new index, the result of the summation remains unchanged.

Formulas (1.3.4) may also be written as follows

$$\bar{e}' = A\bar{e}, \qquad \bar{e} = B\bar{e}', \tag{1.3.5}$$

where $\bar{e}$, $\bar{e}'$ are *columns* consisting of the vectors of the two bases, and A, B are the *matrices* of the coefficients (a_{ij}) and (b_{ij}), respectively.

We would also like to notice that (1.3.4) has the following consequences

$$\bar{e}'_i = \sum_{j,k} a_{ij}b_{jk}\bar{e}'_k, \quad \bar{e}_i = \sum_{j,k} b_{ij}a_{jk}\bar{e}_k,$$

and, since the vectors of a basis are linearly independent, this means that we must have

$$\sum_j a_{ij}b_{jk} = \sum_j b_{ij}a_{jk} = \delta_{ik}, \tag{1.3.6}$$

where δ_{ik} are the so-called *Kronecker symbols* defined as equal to 1 if $i = k$, and equal to 0 if $i \neq k$. The matrix of these symbols is the unit matrix I. Thus, in matrix algebra the relations (1.3.6) become $AB = I$, $BA = I$, and $B = A^{-1}$, $A = B^{-1}$.

Now any vector $\bar{v}$ may be decomposed in the two given bases as follows

$$\bar{v} = \sum_i x_i\bar{e}_i = \sum_j x'_j\bar{e}'_j,$$

and, by using (1.3.4), we obtain

$$\sum_{i,j} x_i b_{ij}\bar{e}'_j = \sum_j x'_j\bar{e}'_j, \quad \sum_i x_i\bar{e}_i = \sum_{i,j} x'_j a_{ji}\bar{e}_i.$$

Since the vectors of a basis are linearly independent, these relations imply:

1.3.11 Proposition. *If two bases of vectors are related by (1.3.4), the coordinates of a vector with respect to these bases are related by the transformation formulas*

$$x'_j = \sum_i b_{ij} x_i, \qquad x_i = \sum_j a_{ji} x'_j. \tag{1.3.7}$$

If the matrix notation is used, formulas (1.3.7) should be written as

$$x' = B^t x, \qquad x = A^t x', \tag{1.3.8}$$

where x, x' are the columns of the coordinates, and A^t, B^t are the transposed matrices.

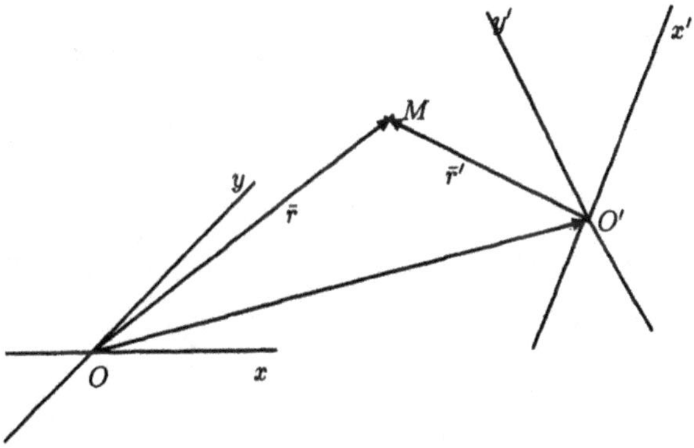

Fig. 1.3.3

Furthermore, for any point M, the radius vectors with respect to the two frames considered earlier are related by (Fig. 1.3.3)

$$\bar r = \vec{OO'} + \bar r'. \tag{1.3.9}$$

Hence, if the coordinates of O' in the first frame are (α_i), we get

$$\sum_j x_j \bar e_j = \sum_j \alpha_j \bar e_j + \sum_i x'_i \bar e'_i,$$

whence, using (1.3.4) and the independence of the vectors $\bar e_i$, we get

1.3.12 Proposition. *The coordinates x_i, x'_i of a point M with respect to two frames are related by means of the following transformation formulas*

$$x_j = \sum_i a_{ij}x'_i + \alpha_j, \quad x'_j = \sum_i b_{ij}x_i + \beta_j, \tag{1.3.10}$$

(where, besides the notation already defined above, β_j are the coordinates of O with respect to the second frame).

In fact, we have proven the first part of (1.3.10) only. But the second part is the same with the roles of the two frames interchanged.

In the matrix notation, and with the notation of formula (1.3.8), we may write (1.3.10) under the form

$$x = A^t x' + \alpha, \qquad x' = B^t x + \beta. \tag{1.3.11}$$

1.3.13 Example. *Consider a cube $ABCDEFGH$ with edges of length 1. Define two frames whose origins are at opposite corners of the cube, e.g., A and G in Fig. 1.3.4, and whose bases are opposite triples of edges starting from the given origins. Write down the change of coordinates of a point when the first frame is replaced by the second.*

Solution. We take the coordinate axes as shown in Fig. 1.3.4. This

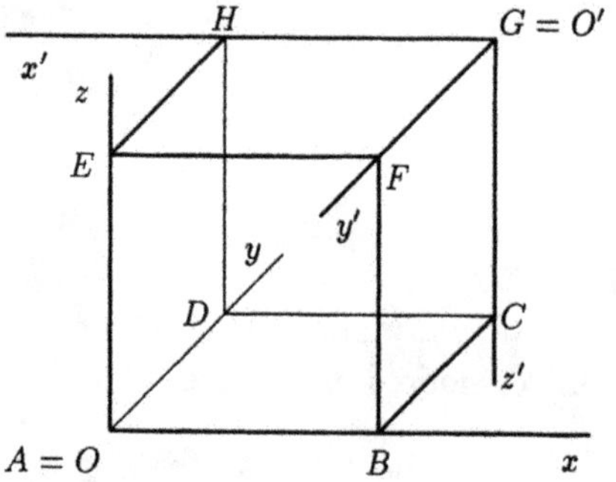

Fig. 1.3.4

means that the two bases are related by

$$\bar{i}' = -\bar{i}, \; \bar{j}' = -\bar{j}, \; \bar{k}' = -\bar{k},$$

and

$$\vec{OO'} = \vec{AG} = \bar{i} + \bar{j} + \bar{k}.$$

Then, if M is an arbitrary point with the coordinates (x, y, z), (x', y', z') with respect to the two frames, respectively, the relation (1.3.9) becomes in our case

$$x\bar{i} + y\bar{j} + z\bar{k} = \bar{i} + \bar{j} + \bar{k} + x'\bar{i}' + y'\bar{j}' + z'\bar{k}',$$

and in view of the relations which exist between the two bases, we get

$$x' = -x + 1, \; y' = -y + 1, \; z' = -z + 1.$$

Exercises and Problems

1.3.1. Draw pictures showing an affine frame and an orthogonal frame, and the points which have the following coordinates: $A(3, -2, 1)$, $B(-2, 1/2, -1/3)$.

1.3.2. Consider the points of coordinates $P(3, -1, 2)$, $M(a, b, c)$, with respect to an orthonormal system of coordinates. Compute the coordinates of the points obtained by a reflection (symmetry) of P, M with respect to the coordinate planes, the coordinate axes and the origin. (Remember that *reflection or symmetry* means to take a perpendicular from the point to the reflecting plane or line, or just to join with the reflecting point, and to continue this perpendicular by the same length on the other part of the reflecting plane, line or point, respectively.)

1.3.3. Consider a regular hexagon $ABCDEF$, and in the plane of this hexagon take a frame with the origin at A, and the basis $\vec{AB}$, $\vec{AC}$. Compute the coordinates of all the vertices of the hexagon with respect to this frame.

1.3.4. Let $Oxyz$ be an orthogonal frame. On the plane (xy) of this frame, place a cube of edge of length a such that the center of the basis of the cube is in O, and the edges of the cube which are parallel to Oz are in the coordinate planes. Compute the coordinates of the vertices of the cube.

1.3.5. Consider a square $ABCD$ of edge of length 1. In the plane of this square, define two frames, one with origin at A and basic vectors

$\vec{AB}$, $\vec{AD}$, and the other with the origin at the center O of the square, and basic vectors $\vec{OB}$, $\vec{OC}$. Write down the formulas of the coordinate transformations of vectors and points, between these two frames.

1.3.6. In space, consider two orthonormal frames with different origins O, O', but such that the endpoints of the basic vectors $(\bar{i}, \bar{i}')$, $(\bar{j}, \bar{j}')$, $(\bar{k}, \bar{k}')$ represented from O, O', respectively, coincide for each pair. Write down the formulas of the coordinate transformations of vectors and points, between these two frames.

1.4 Products of Vectors

Physics, in particular, suggested several ways of multiplying vectors, and these operations proved to be very useful for geometry as well. One such multiplication which comes from the notion of work of a force is defined as follows

1.4.1 Definition. *For any two vectors $\bar{u}$, $\bar{v}$, the scalar number*

$$\bar{u}.\bar{v} = (\bar{u}, \bar{v}) =< \bar{u}, \bar{v} >:= |\bar{u}||\bar{v}| \cos\alpha, \quad (1.4.1)$$

where the first three expressions are notations, and α is the value of the angle between the directions of the two vectors subject to the condition $0 \leq \alpha \leq \pi$, is called the scalar product of $\bar{u}$ and $\bar{v}$.

See Fig. 1.4.1. The name comes from the fact that the result of scalar multiplication is a number. We mentioned three notations of the scalar product which are usual in literature, but we will use the first of them, generally.

An important interpretation of scalar product is provided by *orthogonal projection*. Let d be an axis with direction defined by a vector $\bar{u}$, and let $\bar{v} = \vec{AB}$. Let A', B' be the intersection points of d with planes which are orthogonal to d and pass through A, B, respectively. Then, $\bar{w} = \vec{A'B'}$ is called the *orthogonal projection* of $\bar{v}$ on d, and it is denoted by $pr_d\bar{v}$ or $pr_{\bar{u}}\bar{v}$ (Fig. 1.4.2). The algebraic length $a(pr_d\bar{v})$ is called the measure of the projection, and it is denoted by $mpr_d\bar{v} = mpr_{\bar{u}}\bar{v}$. In plane geometry, the orthogonal planes may be replaced by orthogonal lines. Generally, it is also possible to define projections which are parallel with a certain given plane or direction but, we will only use orthogonal projections here.

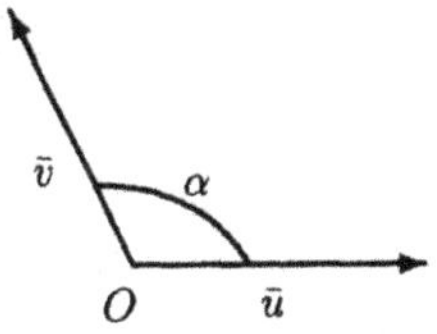

Fig. 1.4.1

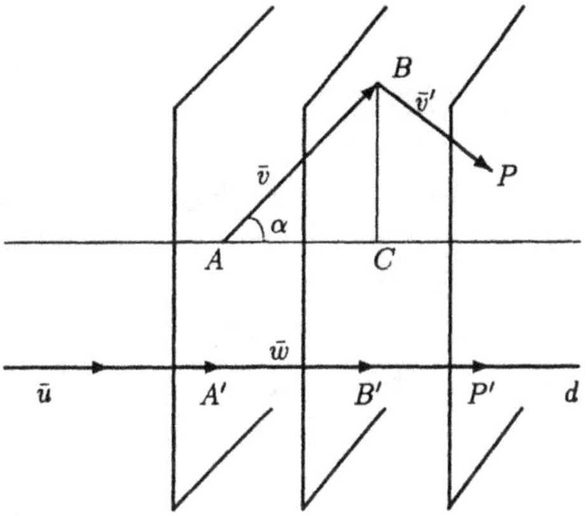

Fig. 1.4.2

As a free vector, $pr_{\bar{u}}\bar{v}$ depends only on the free vectors $\bar{u}$, $\bar{v}$, and it is independent of the concrete choice of d up to parallelism, and of A, B. The same is true for $mpr_{\bar{u}}\bar{v}$, and, from Fig. 1.4.2, we see that

(1.4.2) $$mpr_{\bar{u}}\bar{v} = |\bar{v}|\cos\alpha,$$

where $\alpha \in [0,\pi]$ ($[0,\pi]$ denotes the *interval* of real numbers delimited by 0 and π) is the angle between $\bar{u}$ and $\bar{v}$. Another property which can be seen directly from Fig. 1.4.2 is

(1.4.3) $$pr_{\bar{u}}(\bar{v}+\bar{v}') = pr_{\bar{u}}\bar{v} + pr_{\bar{u}}\bar{v}', \quad mpr_{\bar{u}}(\bar{v}+\bar{v}') = mpr_{\bar{u}}\bar{v} + mpr_{\bar{u}}\bar{v}'.$$

The projection of a vector $\bar{v}$ may be $\bar{0}$ even if $\bar{v} \neq 0$.

The announced interpretation follows from (1.4.2), and it is

(1.4.4) $$\bar{u}.\bar{v} = |\bar{u}|mpr_{\bar{u}}\bar{v} = |\bar{v}|mpr_{\bar{v}}\bar{u}.$$

1.4.2 Proposition. *The basic properties of the scalar product are*

a) $\bar{u}.\bar{v} = \bar{v}.\bar{u}$ *(commutativity);*

b) $\bar{u}.(\bar{v}+\bar{w}) = \bar{u}.\bar{v}+\bar{u}.\bar{w}$ *(distributivity);*
c) $(\lambda\bar{u}).\bar{v} = \bar{u}.(\lambda\bar{v}) = \lambda\bar{u}.\bar{v}$;
d) $\bar{u}^2 := \bar{u}.\bar{u} = |\bar{u}|^2 \geq 0$, *where equality holds iff* $\bar{u} = \bar{0}$;
e) $\bar{u}.\bar{v} = \bar{0}$ *iff the two vectors are perpendicular to one another.*

Proof. *a*), *d*) and *e*) are straightforward consequences of the definition of the scalar product. *b*) is a consequence of (1.4.4) and (1.4.3), and *c*) is checked separately for $\lambda \leq 0$ and $\lambda \geq 0$, by using (1.4.1). Q.E.D.

1.4.3 Remark. *We recall that, in an arbitrary linear space, a scalar product is any operation which satisfies the properties a) – d) of Proposition 1.4.2, and a linear space endowed with a scalar product is a Euclidean linear space.*

Now, let $(O; \bar{i}, \bar{j}, \bar{k})$ be an orthonormal frame. In order to compute scalar products via coordinates, we first use the definition of the scalar product in order to get the scalar product table of the basic vectors:

$$\bar{i}^2 = \bar{j}^2 = \bar{k}^2 = 1, \quad \bar{i}.\bar{j} = \bar{j}.\bar{k} = \bar{k}.\bar{i} = 0. \tag{1.4.5}$$

1.4.4 Proposition. *With respect to orthonormal coordinates, the scalar product of*

$$\bar{u} = x_1\bar{i} + y_1\bar{j} + z_1\bar{k}, \quad \bar{v} = x_2\bar{i} + y_2\bar{j} + z_2\bar{k} \tag{1.4.6}$$

is given by

$$\bar{u}.\bar{v} = x_1x_2 + y_1y_2 + z_1z_2. \tag{1.4.7}$$

Proof. Multiply the coordinate expressions of the two vectors while using the properties of the scalar product (Proposition 1.4.2 *a*), *b*), *c*)), and the multiplication table (1.4.5). Q.E.D.

Of course, along a straight line, the right-hand side of (1.4.7) reduces to its first term, and in plane geometry it reduces to the first two terms only.

The main applications of the scalar product are computations of lengths and angles:

1.4.5 Proposition. *Let* $\bar{u}$, $\bar{v}$ *be vectors given by (1.4.6). Then, we have*

$$|\bar{u}| = \sqrt{\bar{u}^2} = \sqrt{x_1^2 + y_1^2 + z_1^2}, \tag{1.4.8}$$

(1.4.9) $$\cos\alpha = \frac{\bar{u}.\bar{v}}{|\bar{u}||\bar{v}|} = \frac{x_1x_2 + y_1y_2 + z_1z_2}{\sqrt{x_1^2+y_1^2+z_1^2}\sqrt{x_2^2+y_2^2+z_2^2}},$$

where $\alpha \in [0,\pi]$ is the angle between the two vectors. In particular, $\bar{u} \perp \bar{v}$ iff

(1.4.10) $$x_1x_2 + y_1y_2 + z_1z_2 = 0.$$

Proof. The results are straightforward consequences of the definition of the scalar product and of its coordinate expression (1.4.7) in orthonormal coordinates. Q.E.D.

1.4.6 Remark. *With respect to an affine frame $(O;\bar{e}_1,\bar{e}_2,\bar{e}_3)$, if we denote $g_{ij} := \bar{e}_i.\bar{e}_j$, and if $\bar{u} = \sum_i u_i\bar{e}_i$, $\bar{v} = \sum_j v_j\bar{e}_j$, we get*

(1.4.11) $$\bar{u}.\bar{v} = \sum_{i,j} g_{ij}u_iv_j.$$

This formula is more complicated, and we shall not use it in this book. It also leads to some more complicated formulas for the computation of lengths and angles based on the first parts of formulas (1.4.8) and (1.4.9).

1.4.7 Proposition. *Let $(\bar{i},\bar{j},\bar{k})$ be an orthonormal basis of vectors, and let $\bar{u}$ be a unit vector ($|\bar{u}| = 1$) which makes the angles α,β,γ with $(\bar{i},\bar{j},\bar{k})$, respectively. Then*

(1.4.12) $$\bar{u} = (\cos\alpha)\bar{i} + (\cos\beta)\bar{j} + (\cos\gamma)\bar{k}.$$

Proof. We must have a coordinate expression $\bar{u} = u_1\bar{i} + u_2\bar{j} + u_3\bar{k}$, and it leads to $u_1 = \bar{u}.\bar{i} = \cos\alpha$, $u_2 = \bar{u}.\bar{j} = \cos\beta$, $u_3 = \bar{u}.\bar{k} = \cos\gamma$. Q.E.D.

The numbers $(\cos\alpha,\ \cos\beta,\ \cos\gamma)$ are called the *direction cosines* of the direction of the vector $\bar{u}$, and they satisfy the following relation, which follows from $|\bar{u}| = 1$:

(1.4.13) $$\cos^2\alpha + \cos^2\beta + \cos^2\gamma = 1.$$

Generally, if $\bar{u} = u_1\bar{i} + u_2\bar{j} + u_3\bar{k}$, the numbers (u_1,u_2,u_3) are called the *direction parameters* of the unmarked direction of the vector $\bar{u}$. The direction parameters of a direction are only defined up to an arbitrary nonzero factor, since $\bar{u}$ and $\lambda\bar{u}$ $(0 \neq \lambda \in \mathbf{R})$ have the same unmarked direction.

The result of Proposition 1.4.7 is interesting since it shows us how the orthogonal coordinate transformation formulas look like. Namely, they are given by (1.3.7) and (1.3.10) where the lines of the coefficient matrices A, B of (1.3.5) consist of cosines of mutually orthogonal directions. Such matrices are called *orthogonal matrices.* For instance, in plane geometry, if we have two counterclockwise oriented orthonormal bases $(\bar{i}, \bar{j})$ and $(\bar{u}, \bar{v})$, and the angle between $\bar{i}$ and $\bar{u}$ is α (Fig. 1.4.3), the relation between these two bases is

$$\bar{u} = (\cos\alpha)\bar{i} + (\sin\alpha)\bar{j}, \quad \bar{v} = -(\sin\alpha)\bar{i} + (\cos\alpha)\bar{j}. \tag{1.4.14}$$

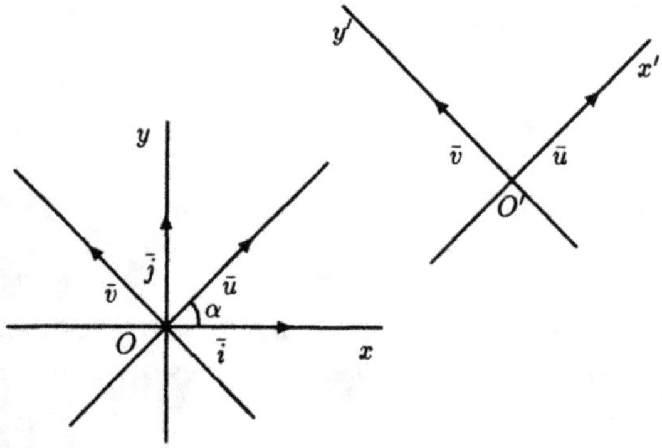

Fig. 1.4.3

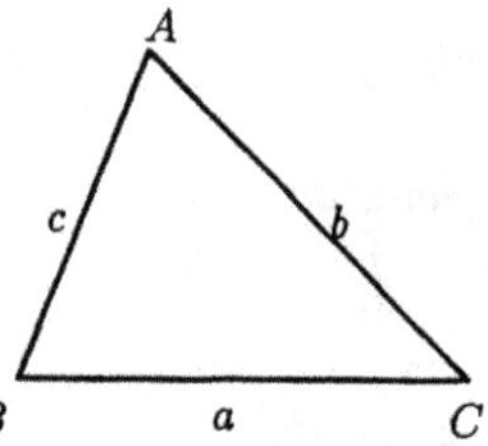

Fig. 1.4.4

Then the corresponding coordinate transformation formulas (1.3.10) will be of the form

$$x' = x\cos\alpha + y\sin\alpha + a, \quad y' = -x\sin\alpha + y\cos\alpha + b, \tag{1.4.15}$$

where (a, b) are the coordinates of O with respect to the second frame (Fig. 1.4.3).

Before leaving this subject, let us do some exercises with scalar products.

1.4.8 Problem. *In an arbitrary triangle ΔABC, prove the cosine theorem: $a^2 = b^2 + c^2 - 2bc\cos\hat{A}$ (Fig. 1.4.4).*

Solution. Notice that $\vec{BC} = \vec{AC} - \vec{AB}$, and compute $\vec{BC}^2$. The result follows simply by using the definition of the scalar product. Q.E.D.

1.4.9 Problem. *Prove that the orthogonal projection of a right angle on a plane α is again a right angle iff at least one of the sides of the given angle is parallel to the plane α.*

Solution. Choose an orthonormal basis $\bar{i}$, $\bar{j}$, $\bar{k}$ such that $\bar{i}$ and $\bar{j}$ are in α, and let the directions of the sides of the given angle be those of the vectors

$$\bar{a} = a_1\bar{i} + a_2\bar{j} + a_3\bar{k}, \quad \bar{b} = b_1\bar{i} + b_2\bar{j} + b_3\bar{k}.$$

The fact that we have a right angle means

$$a_1b_1 + a_2b_2 + a_3b_3 = 0. \tag{1.4.16}$$

Then, the projected angle has the directions of its sides defined by

$$\bar{a}' = a_1\bar{i} + a_2\bar{j}, \quad \bar{b}' = b_1\bar{i} + b_2\bar{j},$$

with the scalar product $\bar{a}'.\bar{b}' = a_1b_1 + a_2b_2$. From (1.4.16), we see that $\bar{a}'.\bar{b}' = 0$ iff $a_3b_3 = 0$. Q.E.D.

Our next operation will be a product of two vectors whose result is a new vector. But we need first some information about the *orientation* of plane and space. In a plane, we are used to consider the clockwise and counterclockwise sense of an angle, which are considered the negative and positive sense, respectively. The existence of positive and negative angles is called the *orientability* of the plane, and the actual choice of the names *positive* and *negative* is called *orientation.* The orientation depends on the observer who looks at the clock in the plane from the unmarked perpendicular direction, and observers who look from two different parts of this unmarked direction see different orientations. (I.e., the positive of one of them will be the negative of the other, and conversely.) Thus, while orientability is a property of the plane, the choice of an orientation is a matter of convention. Orientability is, in fact, the existence of a division of the set of ordered pairs of noncollinear vectors $(\bar{u}, \bar{v})$ of the plane into two disjoint classes, positive and negative pairs according to the sign of the angle $\theta := \pm\alpha$, where $\alpha \in [0, \pi]$ is the angle between $\bar{u}$ and $\bar{v}$, used in the definition of the scalar product, and the sign is determined by the fact that this angle has to be oriented from $\bar{u}$ to $\bar{v}$, the given order of the pair.

In fact, it is possible to define the orientability of the plane without using a clock, which is not an object of the realm of mathematics. Let us sketch how to do this. It is enough to divide the set of orthonormal bases into two classes in order to get an orientation. Fix one such basis $(\bar{\imath}, \bar{\jmath})$, and look at another one $(\bar{\imath}', \bar{\jmath}')$. Rotate $\bar{\imath}'$ by an angle $\leq \pi$ until it reaches $\bar{\imath}$. If this rotation also brings $\bar{\jmath}'$ onto $\bar{\jmath}$, put $(\bar{\imath}', \bar{\jmath}')$ in class No. 1. If not, put it in class No. 2. (Try to fill in the missing details of this proof of the orientability property of the plane.)

In a similar way, *orientability* of space is the property that the ordered triples of noncoplanar vectors $(\bar{u}, \bar{v}, \bar{w})$ may be divided into two disjoint classes by the following convention. If an observer who looks from $\bar{w}$ sees the angle $\alpha \in [0, \pi]$ from $\bar{u}$ to $\bar{v}$ counterclockwise (Fig. 1.4.5), the ordered

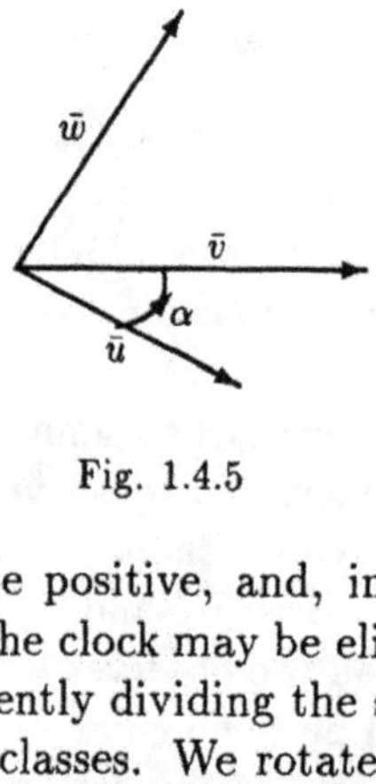

Fig. 1.4.5

triple $(\bar{u}, \bar{v}, \bar{w})$ is said to be positive, and, in the opposite case, this triple is negative. (Again, the clock may be eliminated by reducing the problem to that of conveniently dividing the set of orthonormal bases $(\bar{\imath}, \bar{\jmath}, \bar{k})$, $(\bar{\imath}', \bar{\jmath}', \bar{k}')$ into two classes. We rotate $\bar{\imath}'$ onto $\bar{\imath}$, then compare the pairs $(\bar{\jmath}, \bar{k})$ $(\bar{\jmath}'', \bar{k}'')$, where the latter is the position reached by $(\bar{\jmath}'\bar{k}')$ after the first rotation. These pairs are in the same plane, orthogonal to $\bar{\imath}$, and if they belong to the same class of pairs in this plane, the original triples will be put in the same class. Etc.)

1.4.10 Proposition. *i) If two vectors of a triple are exchanged between them, the sign of the triple is changed. ii) The triples* $(\bar{u}, \bar{v}, \bar{w})$,

$(\bar{v},\bar{w},\bar{u})$, $(\bar{w},\bar{u},\bar{v})$ obtained by cyclic permutations have the same orientation.

Proof. Apply the definition, ... and use your clock!

1.4.11 Definition. *The vector product of two vectors $\bar{u}$, $\bar{v}$ is a vector denoted by $\bar{u}\times\bar{v}$ which satisfies the following conditions: i) its length is numerically equal to the area of the parallelogram defined by the given vectors, ii) it is perpendicular to the given vectors, iii) its direction is such that the triple $(\bar{u},\bar{v},\bar{u}\times\bar{v})$ is positive.*

1.4.12 Proposition. *Let $\bar{n}$ be a unit vector which is orthogonal to $\bar{u}$ and $\bar{v}$, and let θ be the oriented angle from $\bar{u}$ to $\bar{v}$ for an observer who looks from $\bar{n}$ (as defined earlier: $\theta = \pm\alpha \in [0,\pi]$). Then we have*

$$\bar{u}\times\bar{v} = |\bar{u}||\bar{v}|(\sin\theta)\bar{n}. \tag{1.4.17}$$

Proof. We must check that the vector defined by (1.4.17) satisfies the three conditions of Definition 1.4.11. It is clear that the vector (1.4.17) has the necessary direction. As for its length, we get

$$|\bar{u}\times\bar{v}| = |\bar{u}||\bar{v}|\sin\alpha \tag{1.4.18}$$

where $\alpha = |\theta|$, and this is a known elementary formula for the area of a parallelogram. Q.E.D.

Now, we establish the basic properties of the vector product operation.

1.4.13 Proposition. *For vector products, the following properties hold for any vectors and scalars:*
i) $\bar{u}\times\bar{v} = \bar{0}$ iff the vectors $\bar{u}$, $\bar{v}$ are linearly dependent;
ii) anticommutativity or skewcommutativity, which means that

$$\bar{u}\times\bar{v} = -\bar{v}\times\bar{u}; \tag{1.4.19}$$

iii) linearity, which means that

$$(\lambda_1\bar{u}_1+\lambda_2\bar{u}_2)\times\bar{v} = \lambda_1(\bar{u}_1\times\bar{v})+\lambda_2(\bar{u}_2\times\bar{v}); \tag{1.4.20}$$

$$\bar{v}\times(\lambda_1\bar{u}_1+\lambda_2\bar{u}_2) = \lambda_1(\bar{v}\times\bar{u}_1)+\lambda_2(\bar{v}\times\bar{u}_2). \tag{1.4.20'}$$

Proof. i) By (1.4.17), $\bar{u}\times\bar{v}=\bar{0}$ if either $\bar{u}=\bar{0}$ or $\bar{v}=\bar{0}$ or $\sin\theta=0$ and the vectors are collinear, which is exactly the linear dependence of the two vectors. ii) The change of order of the factors changes the sign of the oriented angle θ whence we deduce (1.4.19). iii) First, we establish $\forall\lambda,\bar{u},\bar{v}$ the equalities

$$(\lambda\bar{u})\times\bar{v}=\bar{v}\times(\lambda\bar{v})=\lambda(\bar{u}\times\bar{v}). \tag{1.4.21}$$

These are easily checked for each of the cases $\lambda\le 0$, and $\lambda\ge 0$ by using the formula (1.4.17).

The second step of the proof consists of establishing the *distributivity* property

$$(\bar{u}_1+\bar{u}_2)\times\bar{v}=\bar{u}_1\times\bar{v}+\bar{u}_2\times\bar{v}, \tag{1.4.22}$$

for the case of a vector $\bar{v}$ of length 1.

If $\bar{u}_1,\bar{u}_2$ are collinear, we have, for instance $\bar{u}_2=\lambda\bar{u}_1$, and (1.4.22) is an easy consequence of (1.4.21). But, in the case of linearly independent vectors $\bar{u}_1$, $\bar{u}_2$, the proof is more complicated.

In order to achieve it, we notice that, for a unit vector $\bar{v}$, the vector product $\bar{u}\times\bar{v}$ is given by the following geometric construction (see Fig. 1.4.6): represent $\bar{u}$, $\bar{v}$ by oriented segments from a fixed point O and where the segment $\bar{v}$ is of length 1; take the orthogonal projection of $\bar{u}$ onto the plane α which passes through O and is orthogonal to $\bar{v}$; finally, in α, rotate the obtained projection clockwise (with respect to $\bar{v}$) by a right angle. Indeed, the vector obtained in this way satisfies the conditions of Definition 1.4.11.

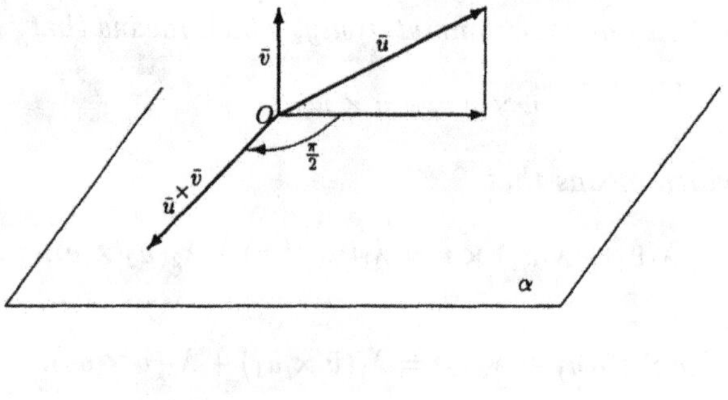

Fig. 1.4.6

Now, assume that $\bar{u}_1$, $\bar{u}_2$ are not collinear. Then, the left-hand side of (1.4.22) is the result of the previous construction of Fig. 1.4.6 applied to the diagonal of the parallelogram defined by $\bar{u}_1$, $\bar{u}_2$. The right-hand side of (1.4.22) is obtained by first applying our construction to $\bar{u}_1$ and $\bar{u}_2$ separately, and, then, taking the diagonal of the parallelogram of the resulting segments. But it is geometrically clear that the final segment will be the same in both cases. Hence, (1.4.22) holds in this case too.

Finally, we see that (1.4.22) also holds for an arbitrary vector $\bar{v}$. Indeed, such a vector is $\bar{v} = |\bar{v}|\bar{w}$, for a unit vector $\bar{w}$, and we have

$$(\bar{u}_1 + \bar{u}_2) \times \bar{v} = |\bar{v}|[(\bar{u}_1 + \bar{u}_2) \times \bar{w}]$$

$$= |\bar{v}|[\bar{u}_1 \times \bar{w} + \bar{u}_2 \times \bar{w}] = \bar{u}_1 \times \bar{v} + \bar{u}_2 \times \bar{v}.$$

If we put together the properties (1.4.21) and (1.4.22), we obviously obtain (1.4.20) (and (1.4.20′) also because of (1.4.19)). Q.E.D.

The next important thing is to establish the coordinate expression of the vector product.

Before doing this, we recall the definitions of *determinants* :

(1.4.23) $$\begin{vmatrix} a & b \\ a' & b' \end{vmatrix} := ab' - ba',$$

(1.4.24) $$\begin{vmatrix} a & b & c \\ a' & b' & c' \\ a'' & b'' & c'' \end{vmatrix} := a\begin{vmatrix} b' & c' \\ b'' & c'' \end{vmatrix} - b\begin{vmatrix} a' & c' \\ a'' & c'' \end{vmatrix} + c\begin{vmatrix} a' & b' \\ a'' & b'' \end{vmatrix},$$

and so on for higher orders. We also extend formulas (1.4.23) and (1.4.24) to the case where the first line of a determinant consists of vectors, and the other lines consist of numbers.

Now, for all the remaining part of this book, and unless something different is said explicitly, we agree to use only *positively oriented orthonormal bases of vectors*, it is orthonormal bases $(\bar{\imath}, \bar{\jmath}, \bar{k})$ which consist of a positive triple of vectors. (Or a positive pair $(\bar{\imath}, \bar{\jmath})$ in the case of a plane. We do not impose the same assumption to the affine bases.) For the vectors of a positive orthonormal basis, we have the following vector multiplication table:

(1.4.25) $$\bar{\imath} \times \bar{\jmath} = \bar{k}, \quad \bar{\jmath} \times \bar{k} = \bar{\imath}, \quad \bar{k} \times \bar{\imath} = \bar{\jmath}.$$

If we use this table, and Proposition 1.4.13, we can compute $\bar{u} \times \bar{v}$ for

$$\bar{u} = u_1\bar{i} + u_2\bar{j} + u_3\bar{k},\ \bar{v} = v_1\bar{i} + v_2\bar{j} + v_3\bar{k},$$

and the result is

1.4.14 Proposition. *The coordinate expression of a vector product with respect to a positive orthonormal basis is*

$$\bar{u} \times \bar{v} = \begin{vmatrix} \bar{i} & \bar{j} & \bar{k} \\ u_1 & u_2 & u_3 \\ v_1 & v_2 & v_3 \end{vmatrix}. \tag{1.4.26}$$

Proof. Following our previous explanations we get

$$\bar{u} \times \bar{v} = (u_2v_3 - u_3v_2)\bar{i} - \tag{1.4.27}$$

$$-(u_1v_3 - u_3v_1)\bar{j} + (u_1v_2 - u_2v_1)\bar{k}.$$

In view of (1.4.24), this is precisely the required result. Q.E.D.

We conclude our discussion of the vector product by two exercises.

1.4.15 Exercise. *Prove the Lagrange identity:*

$$(\bar{u}.\bar{v})^2 + (\bar{u} \times \bar{v})^2 = \bar{u}^2\bar{v}^2, \tag{1.4.28}$$

where $\bar{u}$,$\bar{v}$ are two arbitrary vectors.

Solution. As usual, by the square of a vector we mean the scalar product of the vector with itself. Then (1.4.28) is a straightforward consequence of (1.4.1), (1.4.18) and of the classical identity: $\sin^2\alpha + \cos^2\alpha = 1$. Q.E.D.

1.4.16 Problem. *Find a coordinate expression of the area of a triangle with respect to an orthonormal frame.*

Solution. Consider the triangle $\Delta A_1A_2A_3$, and let $\bar{r}_1, \bar{r}_2, \bar{r}_3$ be the radius vectors of the vertices, respectively. Since a triangle is half of a parallelogram, the definition of a vector product implies

$$Area(\Delta A_1A_2A_3) = \frac{1}{2}|(\bar{r}_2 - \bar{r}_1) \times (\bar{r}_3 - \bar{r}_1)|, \tag{1.4.29}$$

and the reader can obtain easily the corresponding coordinate expression. In particular, if we are in a plane, and if the coordinates of the point A_i are (x_i, y_i) $(i = 1, 2, 3)$, one gets

$$(1.4.30)\quad Area(\Delta A_1A_2A_3) = \frac{1}{2}|(x_2 - x_1)(y_3 - y_1) - (x_3 - x_1)(y_2 - y_1)|$$

$$= \frac{1}{2}\left| \begin{vmatrix} x_1 & y_1 & 1 \\ x_2 & y_2 & 1 \\ x_3 & y_3 & 1 \end{vmatrix} \right|.$$

In (1.4.30) we have the absolute value of the determinant, and the last equality will be checked with the definition (1.4.24) of a third order determinant. It is also usual to say that the determinant (1.4.30) itself is the *oriented area* of the triangle.

1.4.17 Corollary. *The three points A_i $(i = 1, 2, 3)$ of formula (1.4.30) are collinear iff*

$$(1.4.31)\qquad \begin{vmatrix} x_1 & y_1 & 1 \\ x_2 & y_2 & 1 \\ x_3 & y_3 & 1 \end{vmatrix} = 0.$$

The operations of scalar and vector products may be combined in various interesting ways. Thus, we have

1.4.18 Definition. *If $\bar{u}$, $\bar{v}$, $\bar{w}$, are three vectors, then*

$$(1.4.32)\qquad (\bar{u}, \bar{v}, \bar{w}) := \bar{u}.(\bar{v} \times \bar{w})$$

is called a mixed product.

The geometric significance of the mixed product is given by

1.4.19 Proposition. *i) $(\bar{u}, \bar{v}, \bar{w}) = 0$ iff the three vectors are linearly dependent vectors. ii) $(\bar{u}, \bar{v}, \bar{w}) > 0$ iff the triple of involved vectors is positively oriented, and the same product is < 0 iff the triple is negative. iii) For a positive triple, $(\bar{u}, \bar{v}, \bar{w})$ is equal to the volume of the parallelipiped defined by the three vectors (see Fig. 1.4.7).*

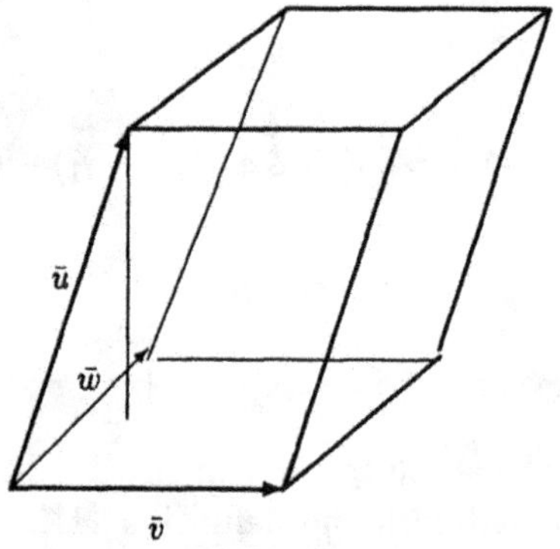

Fig. 1.4.7

Proof. i) By Definition 1.4.18, it follows that the mixed product vanishes iff $\bar{u} \perp \bar{v} \times \bar{w}$, i.e. $\bar{u}$ is in the plane of $\bar{v}$ and $\bar{w}$, which is exactly the linear dependence condition. ii) This is a straightforward consequence of the definition of positively and negatively oriented triples of vectors. iii) By Definition 1.4.18 and by (1.4.4) we have

$$(\bar{u}, \bar{v}, \bar{w}) = |\bar{v} \times \bar{w}| mpr_{\bar{v}\times\bar{w}}\bar{u},$$

which is exactly the area of the basis times the height of the parallelepiped in which we are interested (Fig. 1.4.7). Q.E.D.

In view of iii) above, for any noncoplanar triple, the mixed product $(\bar{u}, \bar{v}, \bar{w})$ is said to be the oriented volume of the associated parallelepiped.

1.4.20 Corollary. *A mixed product is not changed by cyclic permutations of the vectors, and it only changes its sign by the other possible permutations.*

Proof. This is an obvious consequence of Propositions 1.4.10 and 1.4.19: cyclic permutations change neither the orientation nor the volume. Q.E.D.

It is easy to deduce the coordinate expression of the mixed product. Namely,

1.4.21 Proposition. *With respect to a positively oriented orthonormal frame, we have*

$$(1.4.33) \qquad (\bar{u}, \bar{v}, \bar{w}) = \begin{vmatrix} u_1 & u_2 & u_3 \\ v_1 & v_2 & v_3 \\ w_1 & w_2 & w_3 \end{vmatrix},$$

where the lines consist of the coordinates of the three vectors, respectively.

Proof. This follows from (1.4.32) and from the coordinate expressions of the scalar and vector products. Q.E.D.

1.4.22 Problem. *Find a coordinate expression for the calculation of the volume of a tetrahedron with respect to an orthonormal frame.*

Solution. A *tetrahedron* or a pyramid with a triangular basis has four vertices A_i $(i = 1, 2, 3, 4)$ of radius vectors $\bar{r}_i$ and coordinates (x_i, y_i, z_i), respectively. Its volume is one third of the product between the area of the basis $A_1A_2A_3$ and the height from A_4. Therefore, if we double its basis to a parallelogram, the volume will be one sixth of the volume of the parallelipiped defined by the vectors $\bar{r}_2 - \bar{r}_1$, $\bar{r}_3 - \bar{r}_1$ and $\bar{r}_4 - \bar{r}_1$. Hence, we get

$$(1.4.34) \qquad Volume(A_1A_2A_3A_4) = \frac{1}{6}|(\bar{r}_2 - \bar{r}_1, \bar{r}_3 - \bar{r}_1, \bar{r}_4 - \bar{r}_1)|$$

$$= \frac{1}{6}\left| \begin{vmatrix} x_2 - x_1 & y_2 - y_1 & z_2 - z_1 \\ x_3 - x_1 & y_3 - y_1 & z_3 - z_1 \\ x_4 - x_1 & y_4 - y_1 & z_4 - z_1 \end{vmatrix} \right|$$

$$= \frac{1}{6}\left| \begin{vmatrix} x_1 & y_1 & z_1 & 1 \\ x_2 & y_2 & z_2 & 1 \\ x_3 & y_3 & z_3 & 1 \\ x_4 & y_4 & z_4 & 1 \end{vmatrix} \right|.$$

1.4.23 Remark. *The determinants of (1.4.34) are the oriented volume of the tetrehedron. The annulation of these determinants is a necessary and sufficient condition for the four points A_i $(i = 1, 2, 3, 4)$ to be in the same plane.*

1.4.24 Exercise. *Prove that the following identity holds for any three vectors*

$$\bar{u} \times (\bar{v} \times \bar{w}) = (\bar{u}.\bar{w})\bar{v} - (\bar{u}.\bar{v})\bar{w}. \tag{1.4.35}$$

(The left-hand side of formula (1.4.35) is called a double vector product.)

Proof. Decompose $\bar{u}$, $\bar{v}$, $\bar{w}$, with respect to a positive orthonormal basis $\bar{i}$, $\bar{j}$, $\bar{k}$, and use the coordinate expressions of the vector and scalar products. Then (1.4.35) becomes an algebraic identity. Q.E.D.

1.4.25 Exercise. *Prove that the following identity holds for any four vectors*

$$(\bar{v}_1 \times \bar{v}_2).(\bar{v}_3 \times \bar{v}_4) = \begin{vmatrix} \bar{v}_1.\bar{v}_3 & \bar{v}_1.\bar{v}_4 \\ \bar{v}_2.\bar{v}_3 & \bar{v}_2.\bar{v}_4 \end{vmatrix}. \tag{1.4.36}$$

Proof. We have

$$(\bar{v}_1 \times \bar{v}_2).(\bar{v}_3 \times \bar{v}_4) = (\bar{v}_1, \bar{v}_2, \bar{v}_3 \times \bar{v}_4) = \bar{v}_1.(\bar{v}_2 \times (\bar{v}_3 \times \bar{v}_4)).$$

Now, the required identity follows by using the identity (1.4.35) for the double product present here. Q.E.D.

Exercises and Problems

1.4.1. Prove that the heights of any triangle meet at a common point (the *orthocenter of the triangle*).

1.4.2. Prove that the sum of the squares of the lengths of the sides of a parallelogram is equal to the sum of the squares of the lengths of its diagonals.

1.4.3. Prove the *theorem of three perpendiculars*: let α be a plane, d a straight line in α, and O a point outside α; take $OP \perp \alpha$ and $OQ \perp d$; then it follows that $PQ \perp d$.

1.4.4. Prove that if two pairs of opposite edges of a tetrahedron are orthogonal so is also the third pair.

1.4.5. Let $P(-2,1)$, $Q(4,8)$, $R(10,6)$ be three points, their indicated coordinates being with respect to a fixed orthonormal frame. Compute the perimeter of the triangle ΔPQR. Check whether this triangle has any obtuse angle.

1.4.6. In space, choose an orthonormal frame, and consider the points $P(0,-8,-1)$, $Q(3,-2,1)$, and the vector $\bar{v}(10,11,2)$. Compute the length of $\bar{v}$, the distance PQ, and the angle between the vector $\bar{v}$ and the line PQ. Compute the vector obtained by projecting orthogonally $\bar{v}$ onto the axis defined by $\vec{PQ}$.

1.4.7. Let $ABCD$ be a square of side length 2. Write down the coordinate transformation between a plane orthonormal frame whose axes are along adjacent sides of the square $ABCD$, and one whose axes are along the diagonals.

1.4.8. Consider a tetrahedron $OABC$. Let π_A, π_B, π_C be the planes that pass through OA, OB, OC, and are orthogonal to the planes OBC, OCA, OAB, respectively. Show that the planes π_A, π_B, π_C meet along a common straight line.

1.4.9. Let $ABCD$ be a tetrahedron, and $\bar{n}_A$, $\bar{n}_B$, $\bar{n}_C$, $\bar{n}_D$ vectors orthogonal to the opposite faces of A, B, C, D, of length equal to the area of the opposite face, and of direction pointing towards A, B, C, D, respectively. Show that $\bar{n}_A + \bar{n}_B + \bar{n}_C + \bar{n}_D = \bar{0}$.

1.4.10. Let $\bar{i}$, $\bar{j}$, $\bar{k}$, be a positive orthonormal basis, and consider the vectors $\bar{a} = 2\bar{i} + \bar{j} - \bar{k}$, $\bar{b} = \bar{i} - 3\bar{j} + \bar{k}$. Compute the area of the parallelogram defined by the two vectors.

1.4.11. Consider the vectors $\bar{a} = 3\bar{i}+2\bar{j}-5\bar{k}$, $\bar{b} = \bar{i}-\bar{j}+4\bar{k}$, $\bar{c} = \bar{i}-3\bar{j}+\bar{k}$, where $(\bar{i}, \bar{j}, \bar{k})$ is a positive orthonormal basis. Compute the height of their parallelipiped over the basis $(\bar{a}, \bar{b})$.

1.4.12. Let $\bar{a}$, $\bar{b}$, $\bar{c}$, be three noncoplanar vectors. Show that there exists a unique vector $\bar{x}$ which satisfies the system of equations

$$\bar{x}.\bar{a} = \alpha, \ \ \bar{x}.\bar{b} = \beta, \ \ \bar{x}.\bar{c} = \gamma.$$

1.5 A Definition of Affine and Euclidean Space

In the previous sections, our geometric foundations were intuitive. In this section, we indicate a possibility of giving a rigorous mathematical definition of the geometry which we study, and even of more general geometries. This foundation is rather algebraic.

1.5.1 Definition. *A real affine space is a system which consists of a set S of elements called points, and a real vector space V, whose elements will be called vectors, such that the following conditions (axioms) are satisfied:*
i) there exists a unique vector $\bar{v} \in V$ associated with any given, ordered pair of points $A, B \in S$, and we denote this vector by $\bar{v} = \vec{AB}$;
ii) $\forall A \in S$, $\forall \bar{v} \in V$, there exists a unique point B such that the vector $\bar{v}$ is associated to (A, B) in the sense of i);
iii) $\forall A, B, C \in S$, one has $\vec{AB} + \vec{BC} = \vec{AC}$, where the addition of vectors is that of the vectors space structure of V.

The affine space of Definition 1.5.1 is said to be *modeled* on the vector space V, and its study is *affine geometry.* In it, we have the same linear operations with vectors as those encountered in section 1.2, and we can get bases, frames and coordinates of vectors and points, which allows us to develop analytical geometry. But, the space need not be three-dimensional. It may have any number of dimensions, even an infinite one, in agreement with the number of elements of the bases of V.

Furthermore, the Euclidean space (of an arbitrary dimension) is defined by adding an abstract scalar product:

1.5.2 Definition. *An affine space modeled over the linear space V is a Euclidean space if V is also endowed with a scalar product $\bar{u}.\bar{v}$ $(\bar{u}, \bar{v} \in V)$, which axiomatically satisfies the properties a), b), c), d) of Proposition 1.4.2.*

In a Euclidean space, the length of vectors is defined as in property d) of Proposition 1.4.2. Furthermore, we can define as in section 1.4 the distance between two points, the angle of two vectors, etc., and develop the usual geometry of these notions. This is what is called

Euclidean Geometry. Notice, however, that if the space has more than three dimensions, a vector product of two vectors cannot be defined as in the three-dimensional case.

We do not intend to develop here the general abstract affine and Euclidean geometry, as defined in this section, which, therefore, is meant for a general information only. The point is that the utilization of the definitions of this section means to deduce geometry from algebra, while the approach of the previous sections, combined with a full geometric axiomatization of Euclidean space as given by Hilbert e.g. [14], is to deduce algebra from geometry.

Chapter 2

Linear Geometry

2.1 Curves and Surfaces

Starting with this chapter, we study the analytical representations of the simplest curves and surfaces, i.e., the representation of curves and surfaces by means of coordinate systems. We will also study geometric properties of such curves and surfaces by algebraic means, using their coordinate representations.

For the level of this book, it is much too early for a general and rigorous definition of curves and surfaces. Hence, we shall content ourselves with a few intuitive remarks, as far as the general case is concerned.

Following intuition, we conceive a *curve* as a geometric object which corresponds to the form of a thin bent wire, possibly of infinite length (see Fig. 2.1.1). Alternatively, a curve may be seen as the trajectory of a point which moves in a nice way. Accordingly, the coordinates of the moving point with respect to a fixed frame depend on a parameter t (e.g., the *time*), and the curve will be represented by the *parametric equations*:

$$x = f(t), \quad y = g(t), \quad z = h(t), \tag{2.1.1}$$

where f, g, h are functions of t.

The equations (2.1.1) show the basic fact that a curve is a *one-dimensional object* in the sense that *one* independent parameter determines the position of a point of the curve. The nature of this parameter may be arbitrary. For instance, if we can solve the first equation (2.1.1) with respect to t, and get $t = a(x)$, the coordinates of the points of the curve will be characterized by the remaining part of the equations (2.1.1):

$$y = \varphi(x), \quad z = \psi(x), \tag{2.1.2}$$

where $\varphi(x) = g(a(x))$, $\psi(x) = h(a(x))$. Equations (2.1.2) are called *explicit equations* of the curve. (Of course, the roles of the coordinates x, y, z are interchangeable in these equations.)

Sometimes, equations (2.1.2) are equivalent with more general equations of the form

$$F(x, y, z) = 0, \quad G(x, y, z) = 0, \tag{2.1.3}$$

where the passage from (2.1.3) to (2.1.2) is by solving (2.1.3) for y and z. Equations (2.1.3) are called the *implicit equations* of the curve. They define the curve by giving us *characteristic properties* of the points of the curve, i.e., they define the curve as what is called in *elementary geometry* a *geometric locus*. But, of course, not every *locus* in space is a curve; it is a curve only if the points depend on a single independent parameter, in a nice way.

Finally, let us notice that for *plane curves* (i.e., curves which sit in a fixed plane), and if we study problems of plane geometry only, we may use frames whose (x, y)-plane, $z = 0$, is the plane of the curve and then the plane equations of the curve take the following forms:

$$x = f(t), \quad y = g(t) \qquad (\textit{parametric equations}), \tag{2.1.4}$$

$$y = \varphi(x) \qquad (\textit{explicit equation}), \tag{2.1.5}$$

$$F(x, y) = 0 \qquad (\textit{implicit equation}). \tag{2.1.6}$$

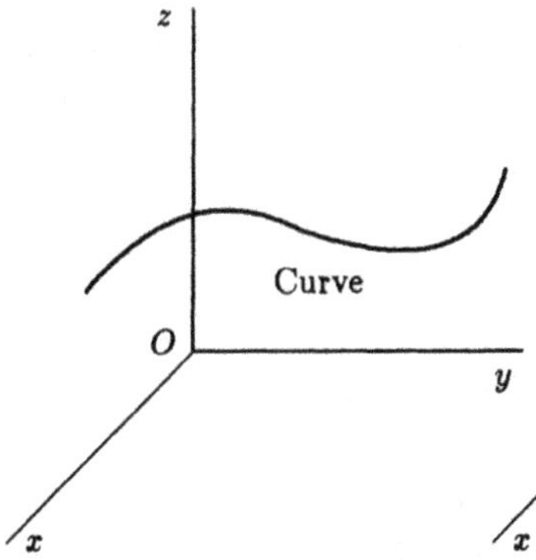

Fig. 2.1.1

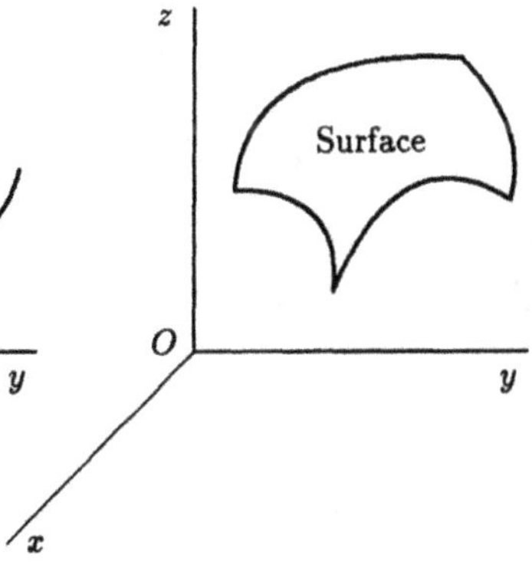

Fig. 2.1.2

Now, again, our intuition tells us that a *surface* should be conceived as a bent, possibly infinite sheet of paper or as the trajectory of a moving curve (Fig. 2.1.2). Since on the curve we already have one parameter, and since, now, the curve moves and its position changes in "time", a point of the surface is determined by the values of two parameters, and the surface is representable by *parametric equations* of the form

$$x = f(u,v), \quad y = g(u,v), \quad z = h(u,v). \tag{2.1.7}$$

In (2.1.7), u, v are *parameters.* The fact that the points of a surface depend on *two* independent parameters is *characteristic* for objects which deserve to be called surfaces and, because of this property we say that a surface is a *two-dimensional* object.

By the same reasoning as for curves, it becomes clear that a surface may also be represented by either one of the following two types of equations:

$$z = \varphi(x,y) \qquad (\textit{explicit equation}), \tag{2.1.8}$$

$$F(x,y,z) = 0 \qquad (\textit{implicit equation}). \tag{2.1.9}$$

It is interesting to compare equations (2.1.9) and (2.1.3). Clearly, the result of this comparison is that a curve also appears as the intersection of two surfaces.

Since this section only has a preliminary and general character, we neither give examples nor do we add exercises now. Such examples and exercises will appear all along the book.

2.2 Equations of Straight Lines and Planes

In this section, we begin the study of the real subject of this chapter. Namely we want to discuss in detail the coordinate representations of straight lines and planes by equations. Since these equations turn out to be of the first degree in x, y, z, i.e., *linear equations*, the geometry of straight lines and planes is called *linear geometry*.

If nothing else is said, we use an affine frame. Clearly, in order to obtain equations for a line or plane we must first define them geometrically, and this may be done in various ways. For instance, we can define a straight line by a point and a direction vector, by two points, etc. Similarly, a plane is defined either by a point and two noncollinear vectors of the plane or by three noncollinear points, etc. Each time, we will obtain corresponding equations, and we begin by

2.2.1 Proposition. *Let d be the straight line which passes through the point M_0, and has the unmarked direction equal to that of the vector $\bar{v}$. Then, the points M of d are given by*

$$\bar{r} = \bar{r}_0 + t\bar{v}, \tag{2.2.1}$$

or, equivalently, by

$$x = x_0 + tv_1, \; y = y_0 + tv_2, \; z = z_0 + tv_3, \tag{2.2.2}$$

where $\bar{r}$, $\bar{r}_0$ are the radius vectors of M and M_0, respectively, $-\infty < t < +\infty$ is a parameter, and the points and vector above have coordinates as follows: $M(x, y, z)$, $M_0(x_0, y_0, z_0)$, $\bar{v}(v_1, v_2, v_3)$.

Proof. The point $M \in d$ iff the vector $\overrightarrow{M_0M}$ is collinear with $\bar{v}$, i.e., $\overrightarrow{M_0M} = t\bar{v}$ (see Fig. 2.2.1). Now (2.2.1) follows from the fundamental formula (1.2.9), and (2.2.2) is just the coordinate expression of (2.2.1). Q.E.D.

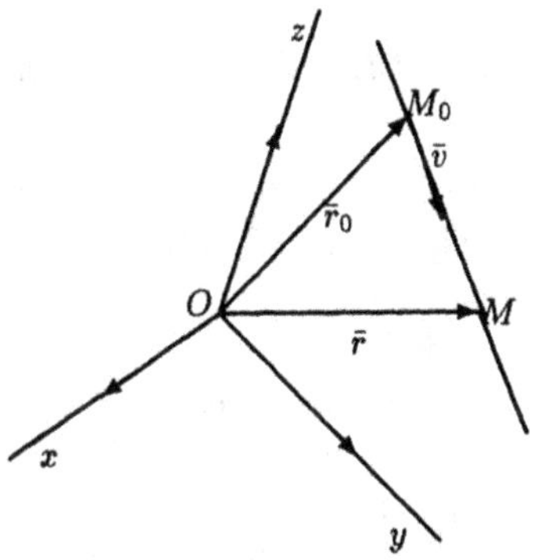

Fig. 2.2.1

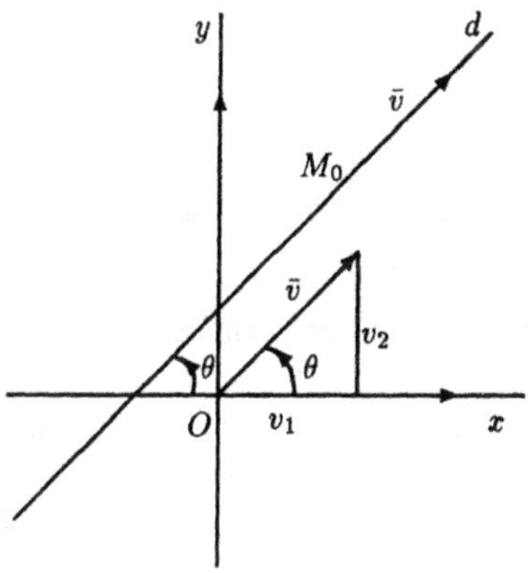

Fig. 2.2.2

(2.2.1) and (2.2.2) are the *vectorial parametric* and *parametric* equations of d, respectively.

2.2.2 Corollary. *i) In plane geometry, (2.2.1) remains unchanged, and (2.2.2) reduces to*

$$x = x_0 + tv_1,\ y = y_0 + tv_2. \tag{2.2.3}$$

ii) The equations of the line d of Proposition 2.2.1 are equivalent to

$$\frac{x - x_0}{v_1} = \frac{y - y_0}{v_2} = \frac{z - z_0}{v_3}. \tag{2.2.4}$$

In plane geometry, only the first equality (2.2.4) remains.
iii) If a straight line d is defined by two points M_1, M_2 of radius vectors $\bar{r}_1(x_1, y_1, z_1)$ and $\bar{r}_2(x_2, y_2, z_2)$, d may be represented by each of the equations

$$\bar{r} = \bar{r}_1 + t(\bar{r}_2 - \bar{r}_1) = t\bar{r}_2 + (1 - t)\bar{r}_1, \tag{2.2.5}$$

$$x = x_1 + t(x_2 - x_1) = tx_2 + (1 - t)x_1,$$

$$y = y_1 + t(y_2 - y_1) = ty_2 + (1 - t)y_1, \tag{2.2.6}$$

$$z = z_1 + t(z_2 - z_1) = tz_2 + (1-t)z_1,$$

(2.2.7) $$\frac{x - x_1}{x_2 - x_1} = \frac{y - y_1}{y_2 - y_1} = \frac{z - z_1}{z_2 - z_1}.$$

In plane geometry, only the first two equations (2.2.6), and only the first equality (2.2.7) remain.

Proof. Each of these results is rather obvious. The terms of (2.2.4) are just the common value of t as given by equations (2.2.2). For the line d of iii), we may ask M_1 to play the role of the point M_0 of the previous cases, and the vector $\bar{v} = \bar{r}_2 - \bar{r}_1$ to define the direction of d. Then, the required equations follow from the previous cases. Q.E.D.

In (2.2.4), (v_1, v_2, v_3) are the *direction parameters* of d, and, in particular, we might use the *direction cosines*, i.e., $v_1^2 + v_2^2 + v_3^2 = 1$ (see Section 1.4). If one of the direction parameters vanishes, the corresponding numerator in (2.2.4) must be annulated, as shown by (2.2.2).

In plane geometry, the equation of d as given by (2.2.4), and if $v_1 \neq 0$, is equivalent to

(2.2.8) $$y - y_0 = m(x - x_0),$$

where $m := v_2/v_1$. m is called the *slope* of d, and, *if the frame is orthonormal*, Fig. 2.2.2 shows that $m = \tan\theta$, where θ is the angle between d, oriented by $\bar{v}$, and the x-axis. If $d \| Oy$ (i.e., $v_1 = 0$), it is common to say that d is of slope $\pm\infty$. If d is determined by two points M_1, M_2, (2.2.7) shows that the slope of d is

(2.2.9) $$m = \frac{y_2 - y_1}{x_2 - x_1}.$$

If $m \neq \pm\infty$, and if M_0 is the intersection point of d with the y-axis, (2.2.8) becomes

(2.2.10) $$y = mx + n,$$

where $n = y_0$. (2.2.10) is the *explicit equation* of a straight line in plane geometry. (Of course, if $d \| Oy$, d has the equation $x = const.$) If d is defined by two points, its equation (2.2.7) is obviously equivalent to

(2.2.11) $$\begin{vmatrix} x - x_1 & y - y_1 \\ x_2 - x_1 & y_2 - y_1 \end{vmatrix} = 0,$$

or to

$$(2.2.12)\qquad \begin{vmatrix} x & y & 1 \\ x_1 & y_1 & 1 \\ x_2 & y_2 & 1 \end{vmatrix} = 0,$$

since an easy computation shows that the determinants of (2.2.11) and (2.2.12) are equal. In orthonormal coordinates, the collinearity condition (2.2.12) was also proven in Corollary 1.4.17. In particular, if d is defined by its intersection points $A(a,0)$, $B(0,b)$ with the coordinate axes, it follows from (2.2.12) that the equation of d may be put under the form

$$(2.2.13)\qquad \frac{x}{a} + \frac{y}{b} - 1 = 0.$$

Coming back to space geometry, we may write again *explicit equations* of a straight line. If we look at (2.2.4), for instance, we see that the line has two independent equations of the first degree, and, assuming that we can solve these equations for y and z, the result will take the form

$$(2.2.14)\qquad y = ax + p, \quad z = bx + q.$$

(2.2.14) are the explicit equations required.

Now, for a plane in space we have

2.2.3 Proposition. *i) The plane α which passes through the point M_0 of radius vector $\bar{r}_0(x_0, y_0, z_0)$, and contains the noncollinear vectors $\bar{u}(u_1, u_2, u_3)$, $\bar{v}(v_1, v_2, v_3)$ is represented by each of the following equations*

$$(2.2.15)\qquad \bar{r} = \bar{r}_0 + t\bar{u} + s\bar{v},$$

$$(2.2.16)\qquad x = x_0 + tu_1 + sv_1,\ y = y_0 + tu_2 + sv_2,\ z = z_0 + tu_3 + sv_3,$$

$$(2.2.17)\qquad \begin{vmatrix} x - x_0 & y - y_0 & z - z_0 \\ u_1 & u_2 & u_3 \\ v_1 & v_2 & v_3 \end{vmatrix} = 0.$$

In these equations $\bar{r}(x, y, z)$ *is an arbitrary point of the plane, and* $s, t \in \mathbf{R}$ *are parameters.*
ii) The plane α *which passes through three noncollinear points* M_i *of radius vectors* $\bar{r}_i(x_i, y_i, z_i)$ $(i = 1, 2, 3)$ *is represented by each of the equations*

$$\bar{r} = \bar{r}_1 + t(\bar{r}_2 - \bar{r}_1) + s(\bar{r}_3 - \bar{r}_1) \tag{2.2.18}$$

$$= s\bar{r}_3 + t\bar{r}_2 + (1 - s - t)\bar{r}_1,$$

$$x = sx_3 + tx_2 + (1 - s - t)x_1,\ y = sy_3 + ty_2 + (1 - s - t)y_1, \tag{2.2.19}$$

$$z = sz_3 + tz_2 + (1 - s - t)z_1,$$

$$\begin{vmatrix} x - x_1 & y - y_1 & z - z_1 \\ x_2 - x_1 & y_2 - y_1 & z_2 - z_1 \\ x_3 - x_1 & y_3 - y_1 & z_3 - z_1 \end{vmatrix} = 0, \tag{2.2.20}$$

$$\begin{vmatrix} x & y & z & 1 \\ x_1 & y_1 & z_1 & 1 \\ x_2 & y_2 & z_2 & 1 \\ x_3 & y_3 & z_3 & 1 \end{vmatrix} = 0. \tag{2.2.21}$$

Proof. Again, to represent a plane (or any other figure) analytically means to establish the characteristic equations satisfied by its points. For i), it is clear that $M \in \alpha$ iff the vectors $\bar{r} - \bar{r}_0$, $\bar{u}$, $\bar{v}$ are linearly dependent vectors (see Fig. 2.2.3). This justifies (2.2.15) and (2.2.16)

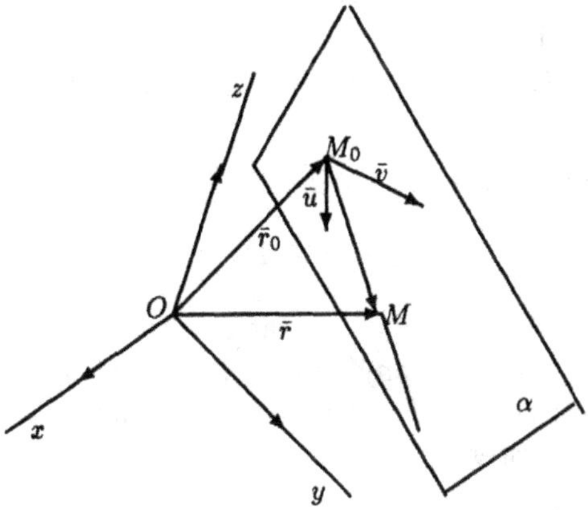

Fig. 2.2.3

in affine coordinates, and (2.2.17) in orthogonal coordinates (because of (1.4.33)). But, (2.2.17) is equivalent to (2.2.16) in affine coordinates as well. Indeed, it suffices to think of the lines of the determinant as orthogonal coordinates of new vectors. The latter will be dependent, and, since linear dependence has an affine character, so will be our vectors too. Then, ii) follows from i) by choosing M_1 as M_0, and $\bar{u} = \bar{r}_2 - \bar{r}_1$, $\bar{v} = \bar{r}_3 - \bar{r}_1$. The coplanarity conditions (2.2.20) and (2.2.21) with respect to orthogonal coordinates were also proven in Remark 1.4.23. Q.E.D.

For instance, the plane which intersects the coordinate axes at the points $A(a,0,0)$, $B(0,b,0)$, $C(0,0,c)$ has the equation

$$\frac{x}{a} + \frac{y}{b} + \frac{z}{c} - 1 = 0. \tag{2.2.22}$$

Indeed, this is what we get from the computation of the corresponding determinant (2.2.20), for instance.

2.2.4 Remark. *The parameters t, s of the parametric equation (2.2.15) of a plane α are affine coordinates in the plane with respect to the frame of origin M_0 and of vector basis $\bar{u}$, $\bar{v}$. Indeed, $\vec{M_0M}$ is exactly the radius vector of M with respect to this frame. We will say that t, s are interior coordinates in α. In particular, if $(\bar{u}, \bar{v})$ is an orthonormal vector basis in α, t, s will be orthonormal interior coordinates.*

Remark 2.2.4 is significant in the study of space geometry of configurations which sit in a certain plane.

Now, we can see the general form of the equations of straight lines and planes:

2.2.5 Proposition. *i) Any plane α has a general equation of the form*

$$ax + by + cz + d = 0, \tag{2.2.23}$$

where a, b, c, d are real constants, and at least one of the coefficients a, b, c is nonzero. Conversely, any equation (2.2.23) where not all the coefficients a, b, c are zero is the equation of a plane. ii) In space, any straight line δ has a system of general equations of the form

$$ax + by + cz + d = 0, \quad a'x + b'y + c'z + d' = 0, \tag{2.2.24}$$

where not all the quotients $a/a', b/b', c/c'$ are equal. Conversely, every such system of equations represents a straight line in space. iii) In plane, every straight line has a general equation of the form

$$ax + by + c = 0, \tag{2.2.25}$$

where a, b are not both zero, and conversely, every such equation represents a straight line with respect to a frame of that plane.

Proof. i) Represent α by equation (2.2.17), and compute the determinant. The result takes the form (2.2.23), and a, b, c are not all zero since the vectors $\bar{u}$, $\bar{v}$ are not proportional. Conversely, if we have an equation (2.2.23) and, say, $c \neq 0$, we may solve it for z, and get an *equivalent explicit equation* of the form

$$z = mx + ny + p \tag{2.2.26}$$

(*equivalent* means that the new equation is satisfied by exactly the same points as the old equation). The following three points satisfy (2.2.26):

$$A(0,0,p), \quad B(1,0,m+p), \quad C(0,1,n+p),$$

and they are noncollinear points (exercise for the reader!). Since the equation of the plane ABC must be a linear equation, it must be equivalent to (2.2.26) hence, it must be equivalent to (2.2.23). Therefore,

equation (2.2.23) represents the plane ABC.
ii) These results follow from i) and from the fact that a straight line δ can always be obtained as the intersection of two nonparallel planes. And, the two planes represented by equations (2.2.24) are parallel iff $a/a' = b/b' = c/c' \neq d/d'$ since, then, the system (2.2.24) has no solutions. Notice also that, iff $a/a' = b/b' = c/c' = d/d'$, the two planes coincide, and they do not define a straight line.
iii) Either argue as for i) or consider the line to be the intersection of the general plane (2.2.23) with the plane $z = 0$. Q.E.D.

2.2.6 Remark. *In space geometry, equation (2.2.25) represents a plane, parallel to the z-axis (why?), not a straight line.*

2.2.7 Proposition. *In space, a vector $\bar{v}(v_1, v_2, v_3)$ belongs to the plane (2.2.23) iff*

$$av_1 + bv_2 + cv_3 = 0. \tag{2.2.27}$$

In plane, $\bar{v}(v_1, v_2)$ belongs to the line (2.2.25) iff

$$av_1 + bv_2 = 0. \tag{2.2.28}$$

Proof. If we put $\bar{v} = \vec{M_1M_2}$ where $M_i(x_i, y_i, z_i)$ $(i = 1, 2)$ belong to the plane, we have

$$v_1 = x_2 - x_1,\ v_2 = y_2 - y_1,\ v_3 = z_2 - z_1,$$

and, by taking the difference of the equalities

$$ax_1 + by_1 + cz_1 + d = 0,\ ax_2 + by_2 + cz_2 + d = 0, \tag{2.2.29}$$

we get (2.2.27). Conversely, if we have (2.2.27), and a point M_1 which satisfies the first condition (2.2.29), the point M_2 of coordinates

$$x_2 = x_1 + v_1,\ y_2 = y_1 + v_2,\ z_2 = z_1 + v_3$$

also satisfies (2.2.29), and it belongs to the given plane. Therefore, $\bar{v}$ belongs to the plane. The case of a straight line in plane is to be treated similarly. Q.E.D.

For the remainder of this section we will use orthonormal coordinates. Then, we can write a few more interesting equations.

2.2.8 Proposition. *In plane, the equation of the straight line which passes through the point M_0 of radius vector $\bar{r}_0(x_0, y_0)$ and is perpendicular to the vector $\bar{n}(n_1, n_2)$ may be written as*

$$\bar{n}.(\bar{r} - \bar{r}_0) = 0 \tag{2.2.30}$$

or, equivalently,

$$n_1(x - x_0) + n_2(y - y_0) = 0, \tag{2.2.31}$$

where $\bar{r}(x, y)$ is the radius vector of an arbitrary point M of the line. In space, (2.2.30) represents the plane through $M_0(x_0, y_0, z_0)$, and orthogonal to $\bar{n}(n_1, n_2, n_3)$, and equation (2.2.31) is replaced by

$$n_1(x - x_0) + n_2(y - y_0) + n_3(z - z_0) = 0. \tag{2.2.32}$$

Proof. Of course, in (2.2.30) the dot is a scalar product. All the results follow from the fact that M belongs to the line (plane) iff $\vec{M_0M} \perp \bar{n}$ (see Fig. 2.2.4). Q.E.D.

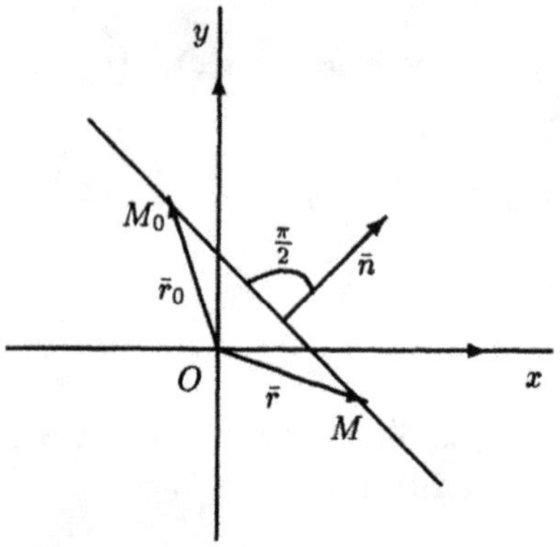

Fig. 2.2.4

2.2.9 Corollary. *With respect to an orthogonal frame, the coefficients a, b, c of the general equation (2.2.23) of a plane (the coefficients a, b of the general equation (2.2.25) of a line in a plane, respectively) are* *coordinates of a vector $\bar{n}$ which is perpendicular to that plane (straight line, respectively).*

Proof. Compare (2.2.23) with (2.2.32) ((2.2.25) with (2.2.31), respectively). Or, notice that Corollary 2.2.9 also follows from the conditions (2.2.27) and (2.2.28) of Proposition 2.2.7 which, now, express the annulation of scalar products, i.e., the orthogonality of (a, b, c) to any vector of the plane (of (a, b) to any vector of the line). Q.E.D.

2.2.10 Definition. *A straight line which is orthogonal to a given plane (straight line) is called the normal of the latter. Similarly, a vector orthogonal to a plane (line) is normal to the latter.*

The reader is asked to keep in mind that the term *normal line* is used in a similar way for all similar geometric situations, i.e., *normal* is used instead of orthogonal whenever we have to speak of a line or a vector which is perpendicular to some interesting plane (line).

In some situations, it is convenient to use equation (2.2.30) for a conveniently oriented, unitary normal vector $\bar{n}$. Then, the coordinates of $\bar{n}$ are $(\cos\alpha, \cos\beta, \cos\gamma)$, where these are the cosines of the normal direction of the plane (respectively, $(\cos\alpha, \sin\alpha)$, where α is the angle of $\bar{n}$ with the x-axis, in plane) (see formulas (1.4.12) and (1.4.14)). The corresponding equations (2.2.31) and (2.2.32) are then

$$x\cos\alpha + y\sin\alpha - p = 0, \tag{2.2.31$'$}$$

$$x\cos\alpha + y\cos\beta + z\cos\gamma - p = 0. \tag{2.2.32$'$}$$

In these equations, p is the *free term*, and the convention is to orient $\bar{n}$ such that $p \geq 0$. This is always possible; just multiply the equation by -1 if necessary. (In some books, the opposite orientation convention is used.) The equations (2.2.31$'$) and (2.2.32$'$) are called the *normal equation* of a line or plane, respectively. It is easy to get the normal equation from the general equations (2.2.25) and (2.2.23): it suffices to take the quotient by the length of the normal vector, i.e., to write

$$\frac{ax + by + c}{\pm\sqrt{a^2 + b^2}} = 0, \tag{2.2.31$''$}$$

$$(2.2.32'') \qquad \frac{ax+by+cz+d}{\pm\sqrt{a^2+b^2+c^2}} = 0,$$

where the sign of the denominator is opposite to the sign of c (d), respectively.

Exercises and Problems

2.2.1. In a plane, consider the lines with the affine (i.e., with respect to an affine frame) equations

$$(d_1)\ 2x - 5y - 1 = 0, \quad (d_2)\ x + 4y - 7 = 0,$$

and the points $A(4,-3)$, $B(-1,2)$. Write down the equation of a straight line which passes through the intersection point of d_1, d_2, and whose intersection point M with the line AB satisfies the condition $(A,B;M) = 2/3$.

2.2.2. Assume that, with respect to an affine frame, the equations of the sides AB, AD of a parallelogram $ABCD$ are

$$(AB)\ x + 2y - 1 = 0 \qquad (AD)\ x + y + 1 = 0,$$

and that the diagonals of $ABCD$ meet at the origin. Write down the equations of the remaining sides of $ABCD$ and of its diagonals. Solve the same problem if

$$(AB)\ ax + by + c = 0 \qquad (AD)\ \alpha x + \beta y + \gamma = 0,$$

and if the diagonals meet at $M(x_0, y_0)$.

2.2.3. In space, and with respect to an orthogonal frame, consider the points $A(1,0,0)$, $B(0,2,0)$, $C(0,0,3)$. i) Write down the equations of the median, the height and the interior bisector of the triangle ΔABC through the vertex A. Write down also the equations of the perpendicular bisector line of BC. ii) Compute the coordinates of the center of gravity, the orthocenter, the center of the circumscribed circle and the center of the inscribed circle of the triangle. iii) Solve question i) above in the general case $A(x_1,y_1,z_1)$, $B(x_2,y_2,z_2)$, $C(x_3,y_3,z_3)$.

2.2.4. In space, consider the lines of affine equations

$$(d_1)\ \frac{x+3}{2} = \frac{y-5}{3} = \frac{z}{1}, \quad (d_2)\ \frac{x-10}{5} = \frac{y+7}{4} = \frac{z}{1}.$$

Write down the equations of a straight line d which meets d_1 and d_2 and is parallel to

$$\frac{x+2}{8} = \frac{y-1}{7} = \frac{z-3}{1}.$$

Compute the coordinates of the intersection points $d \cap d_1$ and $d \cap d_2$. If the frame is orthogonal, compute the direction cosines of the direction of the line d.

2.2.5. Let d be the line of orthogonal (i.e., with respect to an orthogonal frame) equations

$$\frac{x-1}{2} = \frac{y-2}{4} = \frac{z-3}{5},$$

and let P be the point of coordinates $(4, 3, 10)$. Find the coordinates of the symmetric point P' of P with respect to d.

2.2.6. With respect to orthogonal coordinates, consider the points $A(1, -1, 0)$, $B(0, 1, -2)$, $C(-1, 3, 0)$, $D(0, 1, 2)$. Show that $ABCD$ is a rhombus. Write down the equation of the plane of the rhombus and those of its diagonals. Write down also the equations of the straight line d which passes through the intersection point of the diagonals and is perpendicular to both these diagonals. Find the transformation of coordinates between the original frame and an orthonormal frame with the origin at the intersection of the diagonals, and with axes along the diagonals of the rhombus and along the line d.

2.2.7. Let A be one of the vertices of a parallelepiped. Compute the ratio determined on the diagonal through A by the intersection point of this diagonal with the plane which passes through the end vertices of the three sides which begin at A.

2.2.8. A plane π intersects the axes Ox, Oy, Oz of an affine frame at the points A, B, C, respectively. Find the geometric locus of the center of gravity of $\triangle ABC$ if π moves such that i) A is kept fixed, ii)B and C are kept fixed.

2.2.9. With respect to an orthogonal frame, consider the plane $ax + by + cz = 0$, and the vector $\bar{u}(\alpha, \beta, \gamma)$. Compute the coordinates of the orthogonal projection of $\bar{u}$ onto the given plane.

2.3 Geometric Problems on Straight Lines and Planes

Writing down equations of curves or surfaces is not good enough. These equations are to be used in solving geometric problems and, in this section, we discuss a number of such problems concerning straight lines and planes.

Some of these problems are very easy and, in fact, they already appeared in the previous section. For instance, there we deduced conditions for three points to be collinear and for four points to be coplanar. It is also clear how to look for intersection points of lines or planes: we just have to solve the system of equations of those lines and planes taken together, etc.

Some books of Analytical Geometry make a detailed analysis of the possible relative positions of straight lines and planes. Since this is rather easy, we content ourselves with a few sketchy remarks.

In plane, we will classify pairs of straight lines

$$(2.3.1) \qquad (d_1)\ ax + by + c = 0, \qquad (d_2)\ a'x + b'y + c' = 0$$

in accordance to what happens to their intersection. Namely, the lines may coincide, and this happens iff

$$(2.3.2) \qquad \frac{a}{a'} = \frac{b}{b'} = \frac{c}{c'},$$

they may be parallel, and this happens iff

$$(2.3.3) \qquad \frac{a}{a'} = \frac{b}{b'} \neq \frac{c}{c'},$$

or they may intersect at a single point which can be found by solving equations (2.3.1).

Furthermore, three pairwise distinct and nonparallel, coplanar lines either are the sides of a triangle or have a common point whose coordinates are the solutions of (2.3.1) and of the equation $a''x + b''y + c'' = 0$ of the third line. Either a computation or the general theorems on systems of linear equations show that these three equations have a solution

(x, y) iff

$$\begin{vmatrix} a & b & c \\ a' & b' & c' \\ a'' & b'' & c'' \end{vmatrix} = 0. \tag{2.3.4}$$

In a similar way, in space, the pairs of planes

$$(\pi_1) \ \ ax + by + cz + d = 0, \qquad (\pi_2) \ \ a'x + b'y + c'z + d' = 0 \tag{2.3.5}$$

may either be coincident iff

$$\frac{a}{a'} = \frac{b}{b'} = \frac{c}{c'} = \frac{d}{d'} \tag{2.3.6}$$

or parallel iff

$$\frac{a}{a'} = \frac{b}{b'} = \frac{c}{c'} \neq \frac{d}{d'} \tag{2.3.7}$$

or have a straight line as their intersection. Triples of planes "usually" meet at a point but, four planes may have a common point iff

$$\begin{vmatrix} a & b & c & d \\ a' & b' & c' & d' \\ a'' & b'' & c'' & d'' \\ a''' & b''' & c''' & d''' \end{vmatrix} = 0. \tag{2.3.8}$$

Furthermore, two straight lines of space may be either coplanar, which happens iff the four equations of the two lines satisfy condition (2.3.8) (explain why!), or noncoplanar, if they are not parallel and if the four equations have no common solution; we will say that such lines are *skew lines*. If the two lines belong to the same plane then, again they may be either intersecting or parallel or coincident but, of course we must express these facts using the equations of the lines in space. For instance, the lines are parallel iff the direction parameters of the two lines are proportional (why?).

At this stage, it is also simple to compute angles of lines or planes. If d_1, d_2 are two (maybe skew!) straight lines, their unmarked directions define two angles which are either both right angles or one of them is acute and the other is obtuse. Each of them is called an angle between the two lines but, to be precise, one must indicate whether it is the acute or the obtuse angle. Therefore, if $\bar{v}(v_1, v_2, v_3)$, $\bar{w}(w_1, w_2, w_3)$, (where the coordinates, i.e., the direction parameters, are with respect to an orthogonal frame), define the direction of d_1, d_2, respectively, the angle φ of the two lines is given by

$$\cos\varphi = \frac{v_1w_1 + v_2w_2 + v_3w_3}{|\bar{v}||\bar{w}|}, \tag{2.3.9}$$

and this angle is acute if $\cos\varphi > 0$ and obtuse if $\cos\varphi < 0$. Of course, the lines are orthogonal iff the numerator of (2.3.9) vanishes.

If the lines are studied in a plane, it is also possible to give another nice formula for the angle. Namely, if we look at parallels d_1', d_2' to the given lines through the origin, and if we remember the significance of the slope (Fig. 2.2.2), then, by using a well-known trigonometric formula, we get (see Fig. 2.3.1):

$$\tan\varphi = \tan(\theta_2 - \theta_1) = \frac{m_2 - m_1}{1 + m_1m_2}, \tag{2.3.10}$$

where m_1, m_2 are the slopes of the given lines. In particular, $d_1 \perp d_2$ iff $m_1m_2 = -1$.

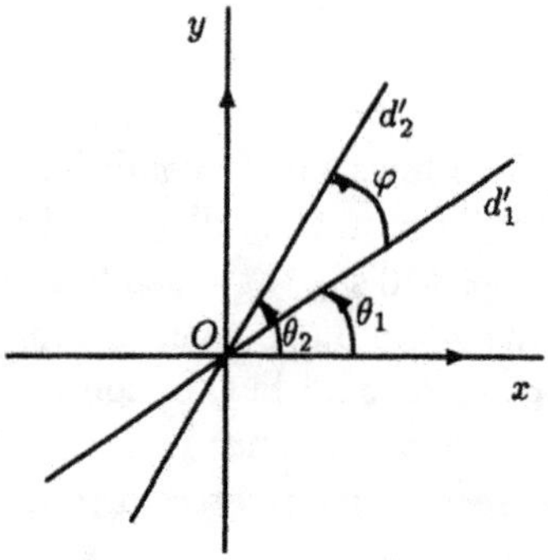

Fig. 2.3.1

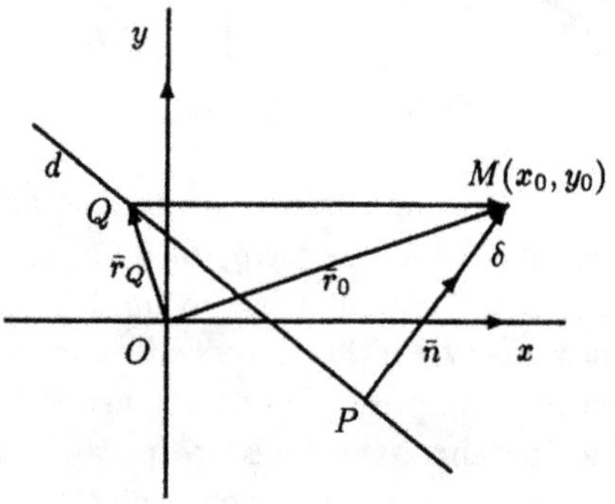

Fig. 2.3.2

The angle between two planes is equal to the angle of their normal directions. Hence, if the planes are given by (2.3.5), and if the coordinates are in an orthogonal frame, we get for the angle φ of these planes:

$$\cos\varphi = \frac{aa' + bb' + cc'}{\sqrt{a^2+b^2+c^2}\sqrt{a'^2+b'^2+c'^2}}. \tag{2.3.11}$$

Of course, if we look at the unmarked normal directions of the planes, we must distinguish between the acute and obtuse angles of the two planes. The planes are perpendicular iff the numerator of (2.3.11) vanishes.

Finally, the angle of a straight line with a plane may be computed as the complement of the angle between the line and the normal of the plane.

Our next problem will be that of computing distances. We will use orthonormal frames, and prove

2.3.1 Proposition. *i) In plane geometry, the distance δ from the point $M_0(x_0, y_0)$ to the line $ax + by + c = 0$ is given by*

$$\delta = \left|\frac{ax_0 + by_0 + c}{\sqrt{a^2+b^2}}\right|. \tag{2.3.12}$$

ii) In space, the distance from $M(x_0, y_0, z_0)$ to the plane $ax+by+cz+d = 0$ is

$$\delta = \left|\frac{ax_0 + by_0 + cz_0 + d}{\sqrt{a^2+b^2+c^2}}\right|. \tag{2.3.13}$$

Proof. The proofs of i) and ii) are the same. To prove i), we represent our data on Fig. 2.3.2, where $\bar{n}$ is a unitary normal vector of the given line d, $MP \perp d$, and Q is an arbitrary point of d with radius vector $\bar{r}_Q$. We see that

$$\delta = |\vec{PM}| = |pr_{\bar{n}}\vec{QM}| = |(\bar{r}_0 - \bar{r}_Q).\bar{n}|$$

(use formula (1.4.4)). Comparing this result with the normal equation (2.2.30), we see that δ is given by the absolute value of the left-hand side of the normal equation of d where the radius vector (hence the coordinates) of the generic point is replaced by that of M. This conclusion, and equation (2.2.31″), prove (2.3.12). Similarly, (2.2.32″) leads to (2.3.13). Q.E.D.

2.3.2 Exercise. Write down the equations of the bisectors of the angles determined by the lines given by (2.3.1), and the equations of the bisector planes of the dihedral angles of the planes (2.3.5).

Solution. Since the bisectors are the locus of the points of equal distance to the sides of the angles, formula (2.3.12) shows that the required bisectors are given by

$$\frac{ax+by+c}{\sqrt{a^2+b^2}} = \pm\frac{a'x+b'y+c'}{\sqrt{a'^2+b'^2}}.$$

The reader is asked to write the similar equation for the bisector planes required in the second part of the exercise.

Notice, that if the intersecting lines d_1, d_2 are given by equations in space, it is more complicated to write equations of the bisectors. For instance, we may represent the bisectors as the intersection of the plane π which contains d_1 and d_2 with the bisector planes of the dihedral angle (α_1, α_2), where α_i $(i = 1, 2)$ is the plane through $M := d_1 \cap d_2$, d_i and the normal of π at M. Otherwise, take two unit vectors $\bar{v}_1$, $\bar{v}_2$ in the direction of d_1, d_2, respectively. Then, the lines through M in the direction of $\bar{v}_1 \pm \bar{v}_2$ are also the required bisector lines (why?).

In space, the distance δ from a point M to a straight line d may be computed by the formula

$$\delta = \frac{|\vec{PM} \times \bar{v}|}{|\bar{v}|}, \tag{2.3.14}$$

where P is an arbitrary point of d, and $\bar{v}$ is a vector which defines the direction of d. This follows if, on Fig. 2.3.3, we notice that δ is the height of the parallelogram $PMSQ$ and, therefore, it is given by the area of the parallelogram divided by the length of its basis PQ.

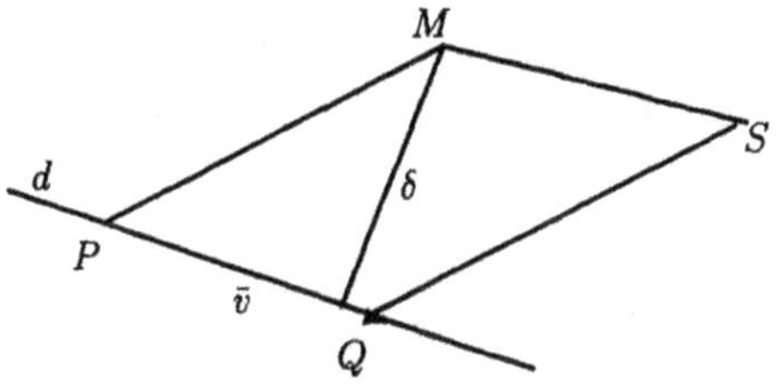

Fig. 2.3.3

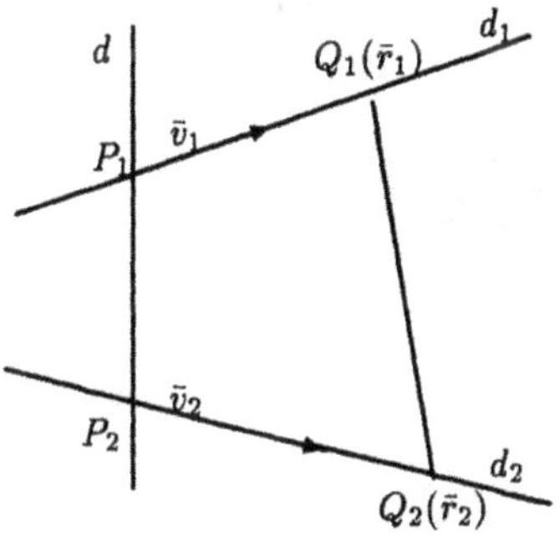

Fig. 2.3.4

In space, again, a distance can also be defined between two non intersecting straight lines d_1, d_2. Namely, a line d which intersects and is orthogonal to both d_1 and d_2 is called a *common perpendicular* of the two lines. If $d \cap d_1 = P_1$ and $d \cap d_2 = P_2$, then the length of P_1P_2 is called the *distance* between the lines d_1, d_2. It is easy to understand (see Fig. 2.3.4) that this distance, say δ, is the length of the smallest segment which joins a point of d_1 and a point of d_2 since the angles formed at P_1, P_2 are right angles. Assuming that the directions of the two lines are defined by the vectors $\bar{v}_1$, $\bar{v}_2$, and that we choose points $Q_1 \in d_1$ and $Q_2 \in d_2$, we immediately get

2.3.3 Proposition. *Two skew straight lines of space have a unique common perpendicular, and, with the notation described above, the distance between the two lines is*

$$\delta = \frac{|(\bar{r}_2 - \bar{r}_1, \bar{v}_1, \bar{v}_2)|}{|\bar{v}_1 \times \bar{v}_2|}. \tag{2.3.15}$$

Proof. Clearly, the direction of the common perpendicular (if it exists) must be that of the vector $\bar{v}_1 \times \bar{v}_2$. Hence, the intersection of the planes α_i determined by d_i and by $\bar{v}_1 \times \bar{v}_2$, for $i = 1, 2$, is the required common perpendicular. The planes are distinct, and the intersection line exists since $\bar{v}_1$, $\bar{v}_2$ are non collinear. The definition of these planes shows that the intersection line is coplanar with both d_1 and d_2, therefore it intersects both of them. The existence of more than one common perpendicular would imply that the given lines are parallel, which is not the case.

Now, we notice (Fig. 2.3.4) that $\vec{P_1P_2}$ is the orthogonal projection of $\vec{Q_1Q_2}$ on $\bar{v}_1 \times \bar{v}_2$. Hence, the algebraic value of this projection is (see formula (1.4.4))

$$(\bar{r}_2 - \bar{r}_1).(\bar{v}_1 \times \bar{v}_2)/|\bar{v}_1 \times \bar{v}_2|.$$

This result obviously implies (2.3.15). Q.E.D.

In a different direction, we have the following

2.3.4 Proposition. *Let M_i $(i = 1, 2)$ be two points of radius vectors $\bar{r}_i$. Then, a point $P(\bar{r})$ belongs to the segment M_1M_2 iff*

$$\bar{r} = \tau\bar{r}_1 + (1 - \tau)\bar{r}_2, \tag{2.3.16}$$

where $0 \le \tau \le 1$.

Proof. We already know that (2.3.16) yields all the points of the straight line M_1M_2; indeed, (2.3.16) is the same as (2.2.5) if t is replaced by $1-\tau$ there. Then, the interior points of the segment M_1M_2 are those with the simple ratio $\kappa = (M_1, M_2; P) > 0$. But, the comparison of (2.3.16) with (1.2.13) shows that $\tau = 1/(1+\kappa)$, therefore, $\kappa = (1-\tau)/\tau$, and $\kappa > 0$ iff $0 < \tau < 1$. Since $\tau = 0$ for M_1, and $\tau = 1$ for M_2, we are done. Q.E.D.

If a formula of the type (2.3.16) or, equivalently $\bar{r} = \lambda\bar{r}_1 + \mu\bar{r}_2$ with $0 \le \lambda, \mu \le 1$, $\lambda + \mu = 1$ holds, one says that the point P is a *convex combination* of M_1, M_2. One may speak of general convex combinations of n points: $\bar{r} = \sum_{i=1}^{n} \lambda_i\bar{r}_i$, where $0 \le \lambda_i \le 1$ and $\sum \lambda_i = 1$.

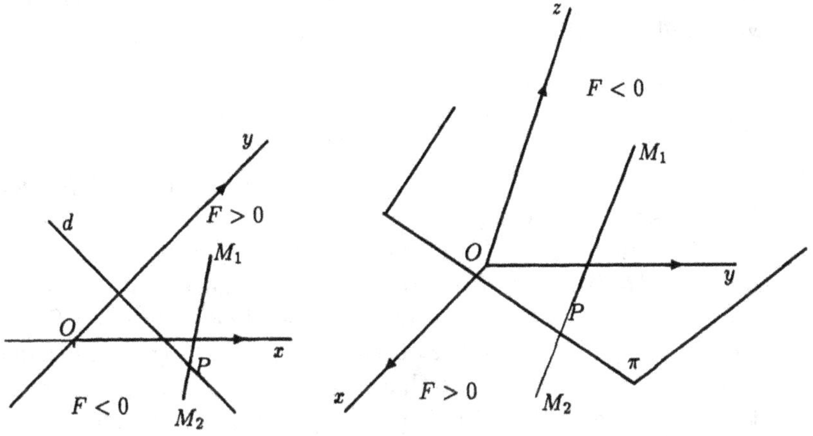

Fig. 2.3.5

It is a geometrically interesting fact that a straight line d of a plane α divides the plane into two parts called *half-planes* such that two points M_1, M_2 belong to different half-planes iff d intersects the segment M_1M_2 at an interior point. Similarly, a plane π divides space into two *half-spaces* where two points M_1, M_2 belong to different half-spaces iff π intersects the segment M_1M_2 at an interior point. (See Fig. 2.3.5.) The analytical expression of this fact is given by

2.3.5 Proposition. *With the notation above, let d have the (affine) equation*

$$F(x,y) := ax + by + c = 0,$$

and let π have the (affine) equation

$$F(x,y,z) := ax + by + cz + d = 0.$$

Then, the points M_1, M_2 belong to different half-planes (half-spaces, respectively) defined by d (π) iff $F(M_1)F(M_2) < 0$, where $F(M_i)$ denotes the value of the expression F for the coordinates of the point M_i ($i = 1, 2$).

Proof. We must express the existence of τ with $0 < \tau < 1$ such that $F(P) = 0$ for the corresponding point P of radius vector $\bar{r}$ given by the convex combination (2.3.16). If we write this combination coordinatewise, and put these coordinates in the expression of F, we easily get

$$F(P) = \tau F(M_1) + (1-\tau)F(M_2) = 0. \tag{2.3.17}$$

Then, it is clear that (2.3.17) has a solution τ as required iff $F(M_1)$ and $F(M_2)$ have opposite signs. Q.E.D.

Of course, since F may be multiplied by an arbitrary number $\lambda \neq 0$, the sign of $F(M_1)$ itself is of no geometric significance. But, if we fix F for d (π), we may characterize the corresponding half-planes (half-spaces) by the sign taken by F at their points. This sign is the same for all the points of a half-plane (half-space), and it changes if we go over to the other half-plane (half-space). (See again Fig. 2.3.5.) For instance, we had the convention that the free term p of the normal equations (2.2.31′), (2.2.32′) of a line (plane) which doesn't contain the origin is positive. Since $-p$ is the value of the left-hand side of these normal equations at the origin, the previous convention meant that we have negative values of the normal form of F in the half-plane (half-space) which contains the origin. With the notation of formula (2.2.30) and of Fig. 2.2.4, we have $p = \bar{n}.\bar{r}_0 = mpr_{\bar{n}}\bar{r}_0$, and we see that $p > 0$ means that $\bar{n}$ is oriented from the origin towards d (π). Furthermore, if we look at Fig. 2.3.2, we see that what may be called the *oriented distance* from a point M to the line (or the plane) namely, the algebraic length $a(\vec{PM})$ (see Section 1.1) is positive in the half-plane (half-space) of positive normal form of F, and negative in the opposite half-plane (half-space).

We will end this chapter by discussing important families of straight lines and planes.

2.3.6 Definition. *A family $\mathcal{F}$ of straight lines in a given plane is a pencil of lines if either i) it consists of all the straight lines through a fixed point M_0, called the base point of the pencil, or ii) it consists of all the straight lines which are parallel to a fixed line d. Similarly, a family of planes in space is a pencil of planes if it is either a family of parallel planes or the family of planes through a base straight line d.*

From this definition, it follows in a straightforward way that a pencil $\mathcal{F}$ of straight lines (planes) has the following properties:
1) two lines (planes) d_1, d_2 in $\mathcal{F}$ determine $\mathcal{F}$ namely, the latter consists of all the lines in the plane of d_1, d_2 (all the planes in space) whose intersection with the given lines (planes) is exactly the same as the intersection of those given lines (planes);
2) there is one and only one line (plane) of the pencil through each point of the plane (space) which is not the base point (does not belong to the base line) of the pencil.

2.3.7 Proposition. *If the straight lines represented by equations (2.3.1) are the lines which determine the pencil $\mathcal{F}$, then the equation*

$$\lambda(ax + by + c) + \mu(a'x + b'y + c') = 0, \tag{2.3.18}$$

where λ, μ are parameters which are not both zero, represents all the lines of $\mathcal{F}$. Similarly, if $\mathcal{F}$ is the pencil of planes defined by the planes (2.3.5), the equation of the pencil $\mathcal{F}$ (i.e., of all its planes) is

$$\lambda(ax + by + cz + d) + \mu(a'x + b'y + c'z + d') = 0, \tag{2.3.19}$$

where, again, λ, μ are parameters which are not both zero.

Proof. If the coefficients of x, y of (2.3.1) are proportional, they are also proportional with the coefficients of x, y of (2.3.18) hence, the latter defines a line which is parallel to the given lines. Otherwise, (2.3.18) passes through the intersection point of the lines (2.3.1) since, obviously, its left-hand side is annihilated by the solution of (2.3.1). Therefore, all the lines which have an equation of the type (2.3.18) belong to $\mathcal{F}$. Conversely, if $d \in \mathcal{F}$, and if $M_0(x_0, y_0) \in d$ is distinct from the base point (if the latter exists), there exists an equation (2.3.18) which defines the line d. Indeed, if we ask that

$$\lambda(ax_0 + by_0 + c) + \mu(a'x_0 + b'y_0 + c') = 0,$$

the pair (λ, μ) is determined up to a common arbitrary factor, and, if this pair is inserted in (2.3.18), we get the required equation of d. The same proof holds for pencils of planes. Q.E.D.

Since it may be inconvenient to work with the pair (λ, μ) because it is determined only up to multiplication with a common factor, we may proceed as follows. Consider all the lines of $\mathcal{F}$ with the exception of the one given by $\lambda = 0$ in (2.3.18). Then, these lines are represented by what is obtained by dividing (2.3.18) by λ, i.e.,

$$ax + by + c + \nu(a'x + b'y + c') = 0, \tag{2.3.20}$$

where $\nu = \mu/\lambda$. Hence, the lines of $\mathcal{F}$ are exactly those defined by (2.3.20), $\forall \nu \in \mathbf{R}$ plus the line $a'x + b'y + c' = 0$. Of course, we can get a similar equation namely,

$$ax + by + cz + d + \nu(a'x + b'y + c'z + d') = 0, \tag{2.3.21}$$

for a pencil of planes.

Here is a typical example of the utilization of pencils:

2.3.8 Exercise. Write down the equation of the plane which passes through the line d of equations

$$\frac{x+2}{1} = \frac{y-3}{2} = \frac{z+1}{-1} \tag{2.3.22}$$

and through the point $A(-1,1,2)$.

Solution. Equations (2.3.22) are equivalent to the system

$$2x - y + 7 = 0, \quad x + z + 3 = 0. \tag{2.3.23}$$

Accordingly, the pencil of planes with base line d consists of the plane α defined by the equation

$$x + z + 3 = 0,$$

and the planes

$$2x - y + 7 + \nu(x + z + 3) = 0. \tag{2.3.24}$$

It is easy to see that A doesn't verify the equation of α. If we ask the coordinates of A to satisfy (2.3.24), we get $4 + 4\nu = 0$, i.e., $\nu = -1$. Inserting this value of ν in (2.3.24) we obtain the equation of the required plane:

$$x - y - z + 4 = 0.$$

In space, one can study complete families of straight lines which are either parallel or have a fixed intersection point. Such families are called *bundles of lines.* Furthermore, in space, a complete family of planes which either have a fixed intersection point or are parallel to a fixed unmarked direction is called a *bundle of planes.* We ask the reader to prove as an exercise that a bundle of planes consists of planes defined by an equation of the form

$$\lambda(ax+by+cz+d)+\mu(a'x+b'y+c'z+d') + \nu(a''x+b''y+c''z+d'')=0, \tag{2.3.25}$$

where $(\lambda, \mu, \nu) \in \mathbf{R}$ are parameters, not all zero, and defined up to a common nonvanishing factor.

More complicated families of straight lines are also of great interest. Namely, the locus of a straight line which moves in space (hence, it depends on the "time" t) is called a *ruled surface*, and the various positions of the moving line are called *generators* of this surface. In particular, if the moving line remains parallel to a fixed direction, the surface is called a *cylinder*, and if the moving line passes through a fixed point M, the surface is a *cone* with *vertex* M. Without going into any detailed study of ruled surfaces, let us only prove

2.3.9 Proposition. *i) In space, an (affine) equation of the form $F(x,y) = 0$ represents a cylinder with generators parallel to the z-axis. ii) An (affine) equation of the form $F(x,y,z)=0$, where the function F is homogeneous (which means that $F(x,y,z)=0$ implies $F(\lambda x, \lambda y, \lambda z)=0$, $\forall\lambda \in \mathbf{R}$) represents a cone with the vertex at the origin.*

Proof. From Section 2.1 we know that the given equations represent surfaces in space. In case i), it is clear that, if $(x_0, y_0, 0)$ is a point of the surface, the same holds for any point with coordinates (x_0, y_0, z) i.e., any point of the parallel to Oz by $(x_0, y_0, 0)$. This shows that the surface is a cylinder as claimed. In case ii), if $M_0(x_0, y_0, z_0)$ is on the surface all the points of the straight line OM_0 belong to the surface because of the homogeneity of F. Hence, the surface is a cone of vertex O. Q.E.D.

2.3.10 Corollary. *i) Any equation of the form*

$$F(ax+by+cz+d, a'x+b'y+c'z+d')=0 \tag{2.3.26}$$

represents a cylinder with generators parallel to the line

$$ax + by + cz + d = 0, \quad a'x + b'y + c'z + d' = 0$$

(if it exists). ii) Any equation of the form

$$F(ax+by+cz+d, a'x+b'y+c'z+d', a''x+b''y+c''z+d'') = 0, \tag{2.3.27}$$

where F is a homogeneous function, represents a cone with the vertex at the intersection point of the three planes defined by the annihilation of the arguments of F (if this point exists and is unique).

Proof. In both cases, a convenient affine change of coordinates brings us back to Proposition 2.3.9. Namely, take

$$x' = ax + by + cz + d, \; y' = a'x + b'y + c'z + d', \; z' = a''x + b''y + c''z + d''.$$

Q.E.D.

Exercises and Problems

2.3.1 In ΔABC, one has $\tan \hat{A} = 1/2$, $\tan \hat{B} = 4/3$, and the equation of the side line AB with respect to some orthogonal frame with the x-axis through A, and the y-axis through B is $2x - y - 2 = 0$. Write down the equations of the side lines AC, BC.

2.3.2. Consider the straight lines

$$4x - 6y - 3 = 0, \quad 2x - 3y + 7 = 0$$

in the (x, y)-plane. Show that they are parallel, and write down the equation of the straight line which is parallel to the given lines and is at equal distance from them.

2.3.3. The vertices B, C of ΔABC are kept fixed, and A moves such that $AB^2 - AC^2 = constant$. Prove that the locus of A is a straight line, and compute the angle of this line with BC.

2.3.4. Find the locus of the centers of the rectangles which are inscribed in a given triangle and have one of their sides on a fixed side of the triangle.

2.3.5. The vertices of ΔABC move along fixed lines OA, OB, OC in such a way that the line AB passes through a fixed point P and the line BC passes through a fixed point Q. Prove that the line AC also passes through some fixed point.

2.3.6. Show that a point P is interior to a triangle $\Delta M_1M_2M_3$ iff the radius vector of P is a convex combination

$$\bar{r} = \sum_{i=1}^{3} \lambda_i \bar{r}_i \quad (0 < \lambda_i < 1)$$

of the radius vectors of the points M_i $(i = 1, 2, 3)$.

2.3.7. i) (Theorem of Menelaos) Consider points $M_1 \in BC$, $M_2 \in AC$, $M_3 \in AB$ (i.e., the points are situated on the corresponding straight lines but, perhaps, outside the indicated segments). Prove that the points M_1, M_2, M_3 are collinear iff

$$(B, C; M_1)(C, A; M_2)(A, B; M_3) = -1.$$

ii) (Theorem of Ceva) In the same configuration as above, prove that the straight lines AM_1, BM_2, CM_3 have a common point iff

$$(B, C; M_1)(C, A; M_2)(A, B; M_3) = 1.$$

2.3.8. In an orthogonal frame of space, write down the equation of a plane which contains the intersection line of

$$x + 5y + z = 0, \quad x - z + 4 = 0,$$

and makes an angle of $\pi/6$ with the plane $x - 4y - 8z + 12 = 0$.

2.3.9. Compute the distance from the point $P(4, 3, -2)$ to the plane $3x - y + 5z + 1 = 0$. (The frame is orthogonal.)

2.3.10. Let d be the intersection line of the planes

$$x + 5y - z + 2 = 0, \quad 4x - y + 3z - 1 = 0.$$

Write down the equation of the plane which passes through d and is: a) parallel to the y-axis, or b) perpendicular to the plane $2x - y + 5z - 3 = 0$.

2.3.11. Consider the straight line d

$$3x + 2y - z + 5 = 0, \quad x - y - z + 1 = 0.$$

(x, y, z are orthogonal coordinates in space.) Write down the equations of the orthogonal projections of d onto the coordinate planes.

2.3.12. Given the points $A(1,1,0)$, $B(0,1,1)$, $C(1,0,1)$, $D(-1,1,1)$, write down the equation of the plane which passes through A, B, and is parallel to the line CD.

2.3.13. Given the points $A(1,-1,1)$, $B(1,1,0)$, $C(0,2,-1)$ (orthogonal coordinates), compute the distance from the point A to the straight line BC.

2.3.14. Write down the equations of a straight line which passes through the point $A(4,0,1)$, and intersects the lines

$$\frac{x-1}{2} = \frac{y+3}{4} = \frac{z-5}{3}, \quad \frac{x}{5} = \frac{y-2}{-1} = \frac{z+1}{2}.$$

2.3.15. Write down the equations of the straight line which passes through the point $A(2,3,1)$, is perpendicular to the line

$$\frac{x+1}{2} = \frac{y}{-1} = \frac{z-2}{3},$$

and it intersects this line.

2.3.16. Let d_1, d_2 be two straight lines with the equations

$$(d_1)\ \frac{x-3}{2} = \frac{y}{-2} = \frac{z-4}{1}, \quad (d_2)\ \frac{x+10}{11} = \frac{y+8}{10} = \frac{z-1}{2}$$

(orthogonal coordinates). i) Show that these lines are coplanar and write down the equation of the plane which they define. ii) Show that one of the angles between d_1 and d_2 is obtuse, and write down the equations of the bisector of this obtuse angle.

2.3.17. Given the straight lines

$$\frac{x+3}{2} = \frac{y-5}{3} = \frac{z}{1}, \quad \frac{x+2}{8} = \frac{y-1}{7} = \frac{z-3}{1}$$

(orthogonal coordinates), write down the equation of their common perpendicular and compute the distance between the two lines.

2.3.18. i) If the edges OA, OB, OC of a tetrahedron are pairwise orthogonal, prove that the planes through the edges OA, OB, OC of a tetrahedron orthogonal to the opposite edges have a common line. ii) Prove that the orthogonal projection of O onto the plane ABC is the orthocenter of the triangle $\triangle ABC$.

Chapter 3

Quadratic Geometry

3.1 Circles and Spheres

This chapter is dedicated to the study of curves and surfaces which have equations of the second degree with respect to affine or orthogonal coordinates, and we begin by the most elementary among them, circles and spheres.

From high school geometry, we remember that the *circle of center A and radius ρ* is the locus of the points M of the plane such that the distance $AM = \rho$. Of course, ρ is a given, real, positive, constant number. The locus of the points M of space such that $AM = \rho$ is called the *sphere of center A and radius ρ*. Since a distance is involved, we will prefer orthogonal coordinates, and we use only such coordinates in the present section. The following results are immediate

3.1.1 Proposition. *i) The circle (sphere) of center A and radius ρ consists of the points M which satisfy the equation*

$$(\bar{r} - \bar{a})^2 = \rho^2, \tag{3.1.1}$$

where $\bar{a}$, $\bar{r}$ are the radius vectors of A, M, respectively, with respect to an arbitrary origin. ii) With respect to orthogonal coordinates in plane, the equation of the circle becomes

$$(x - \alpha)^2 + (y - \beta)^2 = \rho^2, \tag{3.1.2}$$

where α, β are the coordinates of A. Equivalently, (3.1.2) can be written as

$$x^2 + y^2 - 2\alpha x - 2\beta y + \sigma = 0, \tag{3.1.3}$$

where $\sigma = \alpha^2 + \beta^2 - \rho^2$. iii) The similar equations of the sphere with respect to orthogonal coordinates in space are

$$(x-\alpha)^2 + (y-\beta)^2 + (z-\gamma)^2 = \rho^2, \tag{3.1.4}$$

where $A(\alpha, \beta, \gamma)$, and then

$$x^2 + y^2 + z^2 - 2\alpha x - 2\beta y - 2\gamma z + \sigma = 0, \tag{3.1.5}$$

where $\sigma = \alpha^2 + \beta^2 + \gamma^2 - \rho^2$.

Proof. Use formula (1.4.8) to express the distance AM. Then, compute. Q.E.D.

Now, notice that, since an equation of the form (3.1.3) with $\alpha^2+\beta^2-\sigma > 0$ can be put under the form (3.1.2), any such equation represents a circle. For instance, the equation

$$x^2 + y^2 - x + y - 1 = 0 \tag{3.1.6}$$

is the same as

$$(x - \frac{1}{2})^2 + (y + \frac{1}{2})^2 = \frac{3}{2}$$

and it represents the circle of center $(1/2, -1/2)$ and radius $\sqrt{3/2}$. On the other hand, the equation

$$x^2 + y^2 - x + y + 1 = 0 \tag{3.1.7}$$

is the same as

$$(x - \frac{1}{2})^2 + (y + \frac{1}{2})^2 = -\frac{1}{2},$$

and there is no point with real coordinates which satisfies this equation. In both cases, we will say that equation (3.1.3) is the *general equation of a circle* but, in the second case, this is an *imaginary circle*. For the same reason, equation (3.1.5) is *the general equation of a sphere*.

The previous remark shows that a discussion of the geometric objects represented by equations of higher degree must take into consideration the so-called *complex solutions* of these equations. Remember that a *complex number* is an algebraic expression of the form $\lambda+\sqrt{-1}\mu$, where $\sqrt{-1}$ is a symbol supposed to satisfy the condition $(\sqrt{-1})^2=-1$. This symbol is not a real number, since the square of a real number is non-negative, but computations with complex numbers can be done, and they are similar to the computations with real numbers. In particular, complex numbers appear as solutions of *quadratic* (i.e., of degree two) equations in a well-known way. The complex number written above consists of a *real term* λ and a *purely imaginary term* $\sqrt{-1}\mu$ (μ itself is called the *imaginary part* of the number). Accordingly, it is possible to define the *complex geometric plane (space)*, where the points are pairs (triples) of complex coordinates, and it *extends* the *real* plane (space). We do not intend to study *complex geometry* but, we cross it sometimes, as in the case of equation (3.1.7) which represents an *imaginary circle* of radius $\sqrt{-1/2}$. Subsequently, we will keep in mind that such a complex geometry exists.

In the general equation of a circle, there are three parameters (see (3.1.3)). Hence, we may expect that such an equation will be determined by three points which satisfy it. As it is well known from elementary geometry, this happens, indeed, if the given three points are noncollinear, since every triangle has a well-defined circumscribed circle. If the given points are $M_i(x_i, y_i, z_i)$ $(i=1,2,3)$, the equation of the circle is obtained by solving the system of equations

$$x_i^2+y_i^2-2\alpha x_i-2\beta y_i+\sigma=0 \quad (i=1,2,3)$$

with respect to α, β, σ. For the readers who are familiar with determinants it is easy to see that the required circle has the equation

$$\begin{vmatrix} x^2+y^2 & x & y & 1 \\ x_1^2+y_1^2 & x_1 & y_1 & 1 \\ x_2^2+y_2^2 & x_2 & y_2 & 1 \\ x_3^2+y_3^2 & x_3 & y_3 & 1 \end{vmatrix} = 0. \tag{3.1.8}$$

When this determinant is computed, it takes the form (3.1.3) and, on the other hand the determinant vanishes if (x, y) are replaced by (x_i, y_i). (A determinant with two equal lines is zero.)

In exactly the same way, we can use equation (3.1.5) to obtain the equation of a sphere which passes through four noncoplanar points. We leave this task (including the writing down of an equation of the form (3.1.8)) to the reader.

In Section 2.1 we saw that curves and surfaces can also be represented by parametric equations, and we would like to write down such equations for a circle and a sphere. In the case of a circle, if the parameter t is taken to be the angle between the vector $\vec{AM}$ and the x-axis, from Fig. 3.1.1, we immediately get the following values of the coordinates of M:

$$x = \alpha + \rho\cos t,\ y = \beta + \rho\sin t. \tag{3.1.9}$$

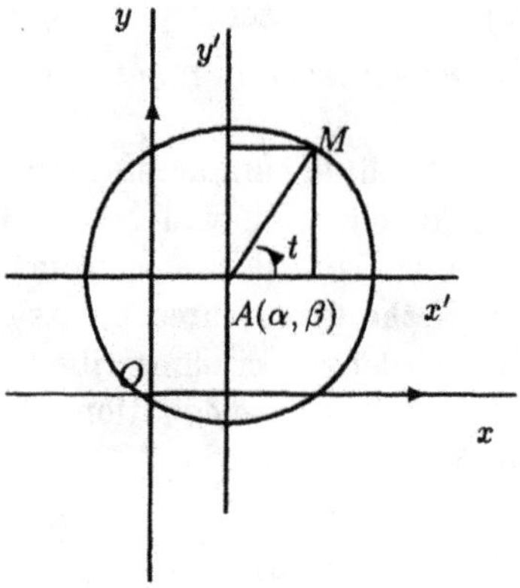

Fig. 3.1.1

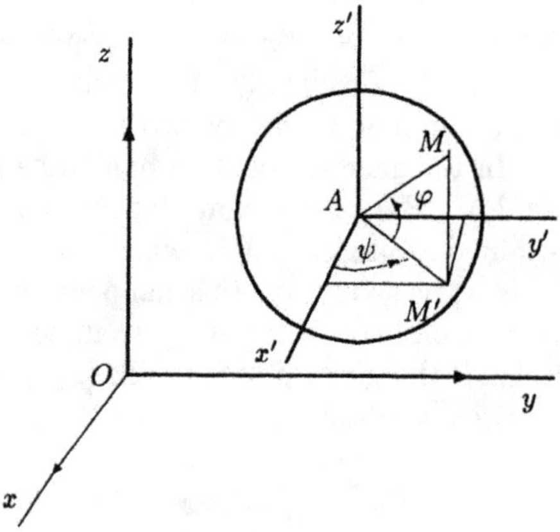

Fig. 3.1.2

Equations (3.1.9) are the *parametric equations of the circle.*

In the case of a sphere, we need two parameters, and we take them to be the *geographic latitude* i.e., the angle φ of $\vec{AM}$ with the plane (x, y), and the *geographic longitude* i.e., the angle ψ between the plane which passes through AM and is parallel to Oz and the plane which passes through A and is parallel to Oxz. Then, from Fig. 3.1.2, we see that the coordinates of M are given by the formulas

$$x = \alpha + \rho \cos\varphi \cos\psi,$$

$$(3.1.10) \qquad y = \beta + \rho \cos\varphi \sin\psi,$$

$$z = \gamma + \rho \sin\varphi,$$

and these are the *parametric equations of the sphere.*

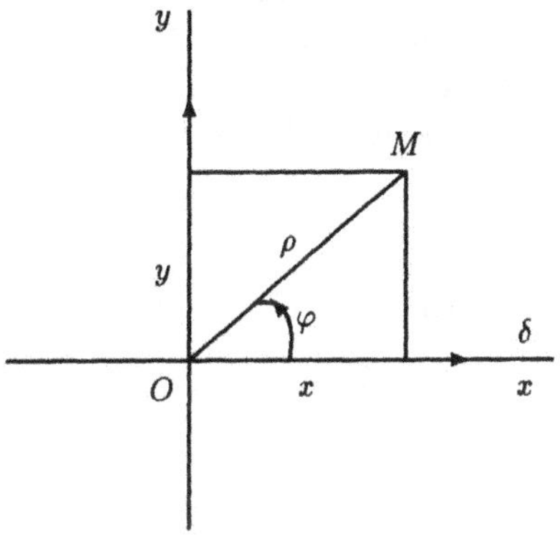

Fig. 3.1.3

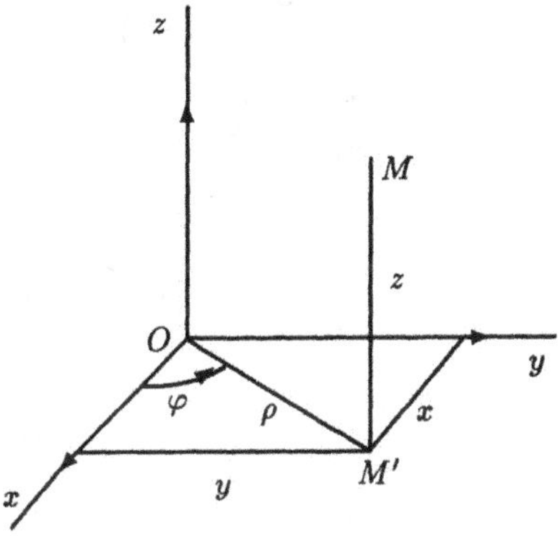

Fig. 3.1.4

Before going on, we want to say that the parameters of the parametric equations of circles and spheres are sometimes used as coordinates of a special kind, and they are important in applications. Namely (see Fig. 3.1.3), in plane geometry, if the *pole* O is fixed, every point M belongs to a well-defined circle of center O, the circle whose radius is $\rho = l(OM)$ (l stands for length). If we also fix an oriented direction δ, which will then be chosen as the direction of Ox, the oriented angle φ from δ to $\vec{OM}$ is also defined, and, conversely, the values (ρ, φ) determine the geometric point M. The pair (O, δ) is called a *polar frame*, and (ρ, φ) are called the *polar coordinates* of M.

There are some problems with the use of polar coordinates. First, these coordinates are not defined for the point O, since φ is not defined for $M = O$ but, we may say that O corresponds to $\rho = 0$. Second, ρ is a length, we must ask $\rho \geq 0$, and the value of the angle φ is determined only up to a term of the form $2n\pi$, where n is an arbitrary integer.

We may fix this ambiguity by asking that $0 \le \varphi < 2\pi$, for instance, but this may lead to certain inconveniences, in particular, if we are interested in a *continuous variation* of φ around 0. Another possibility is to let ρ be an arbitrary real number, even negative, by fixing an orientation of the line OM, and limit φ by $0 \le \varphi < \pi$ (this doesn't fix the discontinuity inconveniences mentioned previously, however). In spite of these problems and inconveniences, polar coordinates are sometimes quite useful.

With a polar frame, we may associate an orthogonal frame namely the frame with the origin at the pole O, the x-axis in the direction of the *polar axis* δ, and the y-axis in the orthogonal direction (making an angle of $+\pi/2$ with δ). From (3.1.9), we see that the transition from the polar coordinates to the associated Cartesian orthogonal coordinate is (see Fig. 3.1.3)

$$x = \rho\cos\varphi, \quad y = \rho\sin\varphi, \tag{3.1.11}$$

and the converse relations are

$$\rho = \sqrt{x^2 + y^2}, \quad \varphi = \arctan\frac{y}{x}. \tag{3.1.12}$$

Similarly, in space, a point M may be associated with a triple of numbers (ρ, φ, z), where (ρ, φ) are the polar coordinates of the projection of M onto the (x, y)-plane, and z is the usual Cartesian height of M (see Fig. 3.1.4). (ρ, φ, z) are called *cylindrical coordinates*, and their relations with the associated Cartesian coordinates are like in (3.1.11) and (3.1.12).

On the other hand, again in space, we may associate with the point M the triple of numbers (ρ, φ, ψ), which consists of the distance $\rho = OM$, and of the latitude φ and the longitude ψ of M on the sphere of center O and radius ρ (see Fig. 3.1.5). These numbers are called the *spherical coordinates* of M, and their relations with the associated Cartesian coordinates are given as in (3.1.10) i.e.,

$$x = \rho\cos\varphi\cos\psi, \ y = \rho\cos\varphi\sin\psi, \ z = \rho\sin\varphi. \tag{3.1.13}$$

Of course, the ambiguity in the values of the angles is also present here. However, the spherical coordinates are very useful in some situations. Let us also notice that if either the latitude, or the longitude, or both are replaced by their complements $\pi/2 - \varphi$, $\pi/2 - \psi$; then, equations (3.1.13) will be changed accordingly; this may be found in some books.

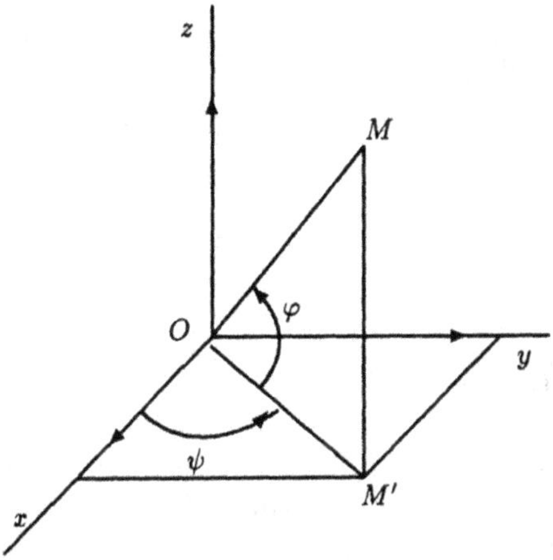

Fig. 3.1.5

Now, we come back to the study of circles and spheres. Many facts about these figures follow from the discussion of their intersection with a straight line. We begin by an analytical proof of the following elementary geometric facts

3.1.2 Proposition. *If Γ is either a circle or a sphere of center A, and d is a straight line in the plane of the circle, the intersection $\Gamma \cap d$ consists of two real points, if the distance from A to d is smaller than the radius of Γ, a double real point, if the distance equals the radius, no real point (two imaginary points), if the same distance is larger than the radius of Γ.*

Proof. We represent Γ by equation (3.1.1), and d by the equation

$$\bar{r} = \bar{r}_0 + \lambda \bar{v}, \tag{3.1.14}$$

where $\bar{r}_0$ is the radius vector of some point M_0 of d, and $\bar{v}$ defines the direction of d. Then, if we insert (3.1.14) into (3.1.1), and if we order the obtained equation following the powers of λ, we get the *intersection equation*

$$\lambda^2\bar{v}^2 + 2\lambda\bar{v}.(\bar{r}_0 - \bar{a}) + (\bar{r}_0 - \bar{a})^2 - \rho^2 = 0. \tag{3.1.15}$$

The required intersection points are obtained by solving the quadratic equation (3.1.15), and inserting the corresponding values of λ in (3.1.14). Hence, we get exactly the three cases mentioned by our proposition, corresponding to $\Delta > 0$, $\Delta = 0$, $\Delta < 0$, where Δ is the so-called discriminant of (3.1.15):

$$\Delta = [\bar{v}.(\bar{r}_0 - \bar{a})]^2 - \bar{v}^2(\bar{r}_0 - \bar{a})^2 + \bar{v}^2\rho^2 = \tag{3.1.16}$$

$$= -|\bar{v} \times (\bar{r}_0 - \bar{a})|^2 + \bar{v}^2\rho^2 = \bar{v}^2[\rho^2 - \frac{|\bar{v} \times (\bar{r}_0 - \bar{a})|^2}{|\bar{v}|^2}].$$

The second equality sign above follows by using the Lagrange identity (1.4.28). But, a comparison with formula (2.3.14) shows that the last term of Δ is just the square of the distance, say h, from the center of Γ to d. Therefore, we have $\Delta = \bar{v}^2[\rho^2 - h^2]$, and $\Delta > 0$ if $h < \rho$, $\Delta = 0$ if $h = \rho$, $\Delta < 0$ if $h > \rho$. Q.E.D.

In the first case (two real intersection points), d is called a *secant.* In the second case, when we have only one intersection point which must be counted twice (a double point), because of the way in which it was obtained algebraically, the line d is called a *tangent of the circle (sphere)*, at the corresponding intersection point (*the contact point*). Since the distance from the center to a tangent line is equal to the radius, we obviously have

3.1.3 Corollary. *Every tangent line is perpendicular to the radius of the circle (sphere) at the contact point.*

The tangent lines play an important geometric role, and it is interesting to write down their equations in various geometric situations. In particular, if M_0 is a fixed point in plane (space), we would like to get all the tangent lines to the circle (sphere) of equation (3.1.1) through M_0. If M of radius vector $\bar{r}$ is an arbitrary point of such a tangent line, this tangent line may be represented by (3.1.14) with the direction vector

$$\bar{v} = \overrightarrow{M_0M} = \bar{r} - \bar{r}_0 \tag{3.1.17}$$

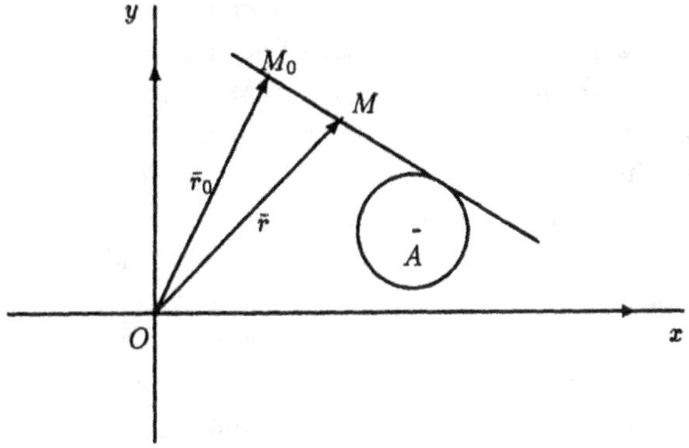

Fig. 3.1.6

(see Fig. 3.1.6). Therefore, the straight line M_0M is tangent to the circle (sphere) iff the vector (3.1.17) satisfies $\Delta = 0$, where Δ is given by the first equality (3.1.16) i.e.,

$$[(\bar{r}-\bar{r}_0).(\bar{r}_0-\bar{a})]^2-(\bar{r}-\bar{r}_0)^2[\rho^2-(\bar{r}_0-\bar{a})^2]=0. \tag{3.1.18}$$

Now, a computation will lead us to

3.1.4 Proposition. *The equation*

$$[(\bar{r}-\bar{a}).(\bar{r}_0-\bar{a})-\rho^2]^2-[(\bar{r}-\bar{a})^2-\rho^2][(\bar{r}_0-\bar{a})^2-\rho^2]=0 \tag{3.1.19}$$

represents all the tangent lines to the circle (sphere) through M_0 together.

Proof. What we mean is that $M(\bar{r})$ belongs to a tangent line as described iff $\bar{r}$ satisfies (3.1.19), and what we have to prove is that (3.1.19) is the same equation as (3.1.18). This follows by replacing $\bar{r}-\bar{r}_0=\bar{r}-\bar{a}+\bar{a}-\bar{r}_0$ everywhere in (3.1.18), then expanding the brackets, making cancellations and comparing with the expanded expression of (3.1.19). The computation is lengthy but technical, and we leave it to the reader. Q.E.D.

The geometric results included in formula (3.1.19) are as follows. First, if M_0 belongs to the circle, (3.1.19) reduces to

$$(\bar{r}-\bar{a}).(\bar{r}_0-\bar{a})-\rho^2=0. \tag{3.1.20}$$

(This equation should be counted twice from the algebraic viewpoint, but we are not interested in this aspect.) Clearly, (3.1.20) is just the equation of a straight line, and this line is the *tangent line* of the circle at M_0. It is orthogonal to the radius AM_0 of the circle, which is why we will also refer to the line AM_0 as the *normal of the circle* at M_0. In the case of a sphere, we look at (3.1.20) from the viewpoint of space. Hence, it represents a plane through M_0 and perpendicular to the radius. This plane consists of all the tangents of the sphere at M_0, and we will call it the *tangent plane of the sphere* at the *contact point* M_0. The line of the radius AM_0 is the *normal of the sphere* at M_0.

Formally, we may notice that the left hand side of (3.1.20) is obtained from the left hand side of the equation of the circle (sphere) by replacing the square $(\bar{r}-\bar{a})^2$ by a product where, in one of the factors, $\bar{r}$ is replaced by $\bar{r}_0$. This formal procedure is known in algebra as a *polarization*. If equations (3.1.1) and (3.1.20) are written by means of coordinates, we see that the *polarization* of the general equations (3.1.3), (3.1.5) of a circle (sphere) consists of replacing

$$x^2 \mapsto xx_0, \quad x \mapsto \frac{1}{2}(x+x_0), \tag{3.1.21}$$

and similarly for y and z, where (x_0, y_0, z_0) are the coordinates of M_0. The conclusion is that *the equation of the tangent line (plane) at* $M_0 \in \Gamma$*, where* Γ *is a circle (sphere) is given by polarizing the general equation of* Γ.

3.1.5 Example. Check that the point $M_0(2,-1,3)$ belongs to the sphere Γ of equation

$$x^2+y^2+z^2-x+y+z-14=0,$$

and write down the equation of the tangent plane of Γ at M_0.

Solution. To check that M_0 satisfies the equation of Γ is straightforward. The equation of the tangent plane is obtained by polarization, which gives

$$2x-y+3z-\frac{1}{2}(x+2)+\frac{1}{2}(y-1)+\frac{1}{2}(z+3)-14=0,$$

i.e.,

$$3x-y+7z-28=0.$$

Let us come back to equation (3.1.19) for a point M_0 and a circle (sphere) Γ. If M_0 lies outside Γ, it is geometrically clear that we will have two tangent lines, in the case of the circle, and a *tangent cone*, in the case of a sphere. Equation (3.1.19) represents this pair of lines or the cone. If M_0 is inside Γ, we have no real tangent lines from M_0 to Γ, but one can say that equation (3.1.19) represents a pair of imaginary tangent lines and an imaginary tangent cone, respectively, if we accept to do geometry over the complex numbers as already mentioned earlier.

It is interesting to notice that, since the contact points of the tangents of the circle (sphere) through M_0 must satisfy both equations (3.1.19) and (3.1.1), they must also satisfy (3.1.20). Hence, the line (plane) (3.1.20) has an interesting geometric role for a general point M_0 also. This line (plane) is called the *polar line (plane)* of M_0 with respect to the circle (sphere), and M_0 is the *pole* of the corresponding line (plane). The correspondence point $\mapsto$ straight line (plane) which sends the *pole* to its *polar line (plane)* is very interesting for some geometrical aspects which are beyond our scope.

3.1.6 Example. Find the pole of the plane

$$x + y + z + 1 = 0$$

with respect to the sphere

$$x^2 + y^2 + z^2 - x + y + z - 14 = 0.$$

Solution. Assume that the required pole is $M_0(x_0, y_0, z_0)$. Then, the polar plane is obtained by polarization i.e., the polar plane is

$$xx_0 + yy_0 + zz_0 - \frac{1}{2}(x + x_0) + \frac{1}{2}(y + y_0) + \frac{1}{2}(z + z_0) - 14 = 0,$$

or, equivalently

$$x(x_0 - \frac{1}{2}) + y(y_0 + \frac{1}{2}) + z(z_0 + \frac{1}{2}) + \frac{1}{2}(-x_0 + y_0 + z_0 - 28) = 0.$$

Since this latter equation also represents the given plane, the coefficients must be proportional, and we get

$$x_0 = -\frac{63}{2}, \; y_0 = z_0 = -\frac{127}{4}.$$

Another problem related to tangents of a circle (sphere) is that of writing down the equation of the tangents of a given unmarked direction defined by a fixed vector $\bar{v}$. We will again use the condition $\Delta = 0$ with Δ given by (3.1.16). Only that, now, we let M_0 to be an arbitrary point M of radius vector $\bar{r}$. Hence, the required equation is

$$[\bar{v}.(\bar{r}-\bar{a})]^2 - \bar{v}^2[(\bar{r}-\bar{a})^2 - \rho^2] = 0. \tag{3.1.22}$$

It is geometrically clear that, for a circle, we get two tangents of direction $\bar{v}$, and, in the case of a sphere, we get a circumscribed cylinder with the generators in the direction of $\bar{v}$ (see Fig. 3.1.7).

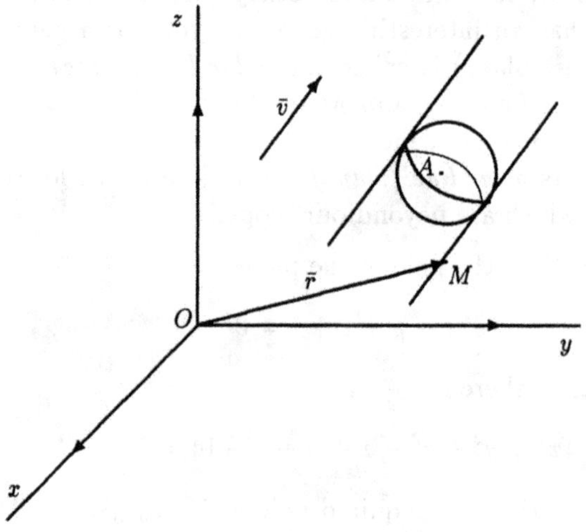

Fig. 3.1.7

The study of the intersection with a straight line also leads to the important concept of the *power* of a point with respect to a circle (sphere).

3.1.7 Proposition. *Let Γ be a circle (sphere) with the general equation (3.1.3) ((3.1.5), respectively), and let M be an arbitrary point in the plane (space) of Γ. Then, if P_1, P_2 are the intersection points of Γ with an axis d through M, the product $a(\vec{MP_1}).a(\vec{MP_2})$ is independent of d.*

Proof. If we represent d by equation (3.1.14), where $\bar{v}$ is a unitary vector, the term of λ^2 of the intersection equation (3.1.15) has the coefficient 1, and $a(\vec{MP_1})$, $a(\vec{MP_2})$ are the solutions λ_1, λ_2 of this intersection equation. Hence, in view of the well-known relations between the coefficients and the solutions of a quadratic equation, the product we want is

$$\lambda_1.\lambda_2 = (\bar{r}_0 - \bar{a})^2 - \rho^2, \tag{3.1.23}$$

where $\bar{r}_0$ is the radius vector of M. This result is independent of d. Q.E.D.

3.1.8 Definition. *The number*

$$p(M, \Gamma) := a(\vec{MP_1}).a(\vec{MP_2})$$

is called the power of M with respect to Γ.

Notice that the intersection points P_1, P_2 of this definition may also be imaginary points if d has no real intersection points with Γ. Furthermore, the result (3.1.23) yields

3.1.9 Proposition. *$p(M, \Gamma)$ is equal to the value of the left hand side of the general equation of Γ with (x, y, z) replaced by the coordinates of M.*

Let us indicate some interesting consequences of Proposition 3.1.9. First, it is geometrically clear that a circle (sphere) Γ separates between two regions of the plane (space), the *interior* and the *exterior* of Γ, and that $p(M, \Gamma) < 0$ iff M is interior to Γ, while $p(M, \Gamma) > 0$ iff M is exterior to Γ. *Hence, the two regions are distinguished between them by the sign of the left hand side of the general equation of* Γ.

Second, let us consider two circles of general equations

$$(\Gamma_1) \qquad x^2 + y^2 - 2\alpha_1 x - 2\beta_1 y + \sigma_1 = 0,$$

$$(\Gamma_2) \qquad x^2 + y^2 - 2\alpha_2 x - 2\beta_2 y + \sigma_2 = 0,$$

and look for the locus of the points of equal power with respect to them. Obviously, the equation of this locus is given by equating the left hand sides of the equations of Γ_1 and Γ_2 i.e.,

$$2(\alpha_2 - \alpha_1)x + 2(\beta_2 - \beta_1)y - (\sigma_2 - \sigma_1) = 0.$$

This is a straight line which passes through the (real or imaginary) intersection points of the two circles, and this line is called the *radical axis* of the two circles. Moreover, we see that the intersection points of two circles are exactly the intersection points of any one of these circles with the radical axis whence, two circles have two intersection points which may either be real and different or real and coincident or imaginary points. Furthermore, if we have three circles, we can group them into three pairs, and the three corresponding radical axes meet (if at all) where the power with respect to the three circles are equal. Usually, this intersection is a unique point, called the *radical center* of the three circles.

Using the general equation (3.1.5) of spheres, the reader will similarly deduce that two spheres have a *radical plane* (which is the locus of the points of equal power with respect to the two spheres, and it passes through all the intersection points of the spheres), three spheres usually have a *radical axis* (which is the locus of the points of equal power with respect to the spheres), and four spheres usually have a point of equal power, called the *radical center* of the spheres.

One of the above remarks raises the problem of studying the intersection of a sphere with a plane, which is important in its own right. Let Γ be the sphere of equation (3.1.1), and let π be a plane defined by a point M_0 of radius vector $\bar{r}_0$ and by two vectors $\bar{v}_1, \bar{v}_2$. It is convenient to choose these vectors of length 1 and perpendicular, such that they yield an *interior* orthogonal frame of π. Then, $M \in \pi$ has the radius vector

$$\bar{r} = \bar{r}_0 + \xi\bar{v}_1 + \eta\bar{v}_2,$$

where (ξ, η) are orthogonal coordinates *in the plane* π. Thus, the intersection points are characterized by the equation obtained if this $\bar{r}$ is inserted into (3.1.1). The result is

$$(3.1.24)\quad \xi^2 + \eta^2 + 2\bar{v}_1.(\bar{r}_0 - \bar{a})\xi + 2\bar{v}_2.(\bar{r}_0 - \bar{a})\eta + (\bar{r}_0 - \bar{a})^2 - \rho^2 = 0,$$

and it is clear that this is the *interior equation* of a circle in the plane π (compare with equation (3.1.3)). Moreover, (3.1.24) shows that the interior coordinates of the center of this circle with respect to the frame $(M_0, \bar{v}_1, \bar{v}_2)$ are exactly those of the orthogonal projection of the center A of the sphere Γ onto π (see Fig. 3.1.8), and the radius of the circle can be computed accordingly. Notice that these calculations give us a method of studying *circles in space.*

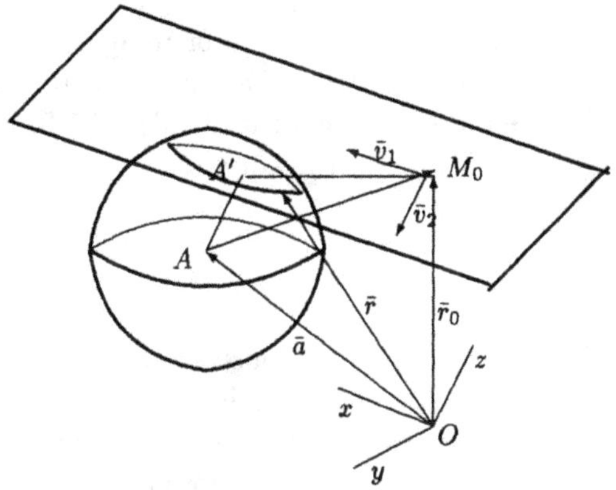

Fig. 3.1.8

Now, it also becomes clear that the intersection of two spheres is again a circle, because it is equal to the intersection of each of the spheres with the radical plane of the two spheres.

If M_0 is an intersection point of two circles (spheres) Γ_1, Γ_2, the *angle* between Γ_1 and Γ_2 at M_0 is defined as the angle between either the tangents (tangent planes) or the normals (i.e., the radii) of the two circles at M_0. (It is clear that the use of either the tangents or the normals gives the same value of the angles; the angles themselves are just rotated by $\pi/2$.) For reasons of symmetry, the angle defined above does not depend on the choice of the intersection point M_0, and, therefore, we may simply speak of the angle of two circles (spheres).

The definition of the angle as given above may be generalized to any two curves or surfaces but, usually, the angle will depend on the intersection point.

Like in the case of straight lines and planes, a notion of a *pencil of circles or of spheres* is very useful, and we define it as follows. A *pencil of circles (spheres)* is the family of all the circles in a plane (spheres in space) which pass through the intersection of either two fixed circles in the plane (spheres) or the intersection of a circle (sphere) and a straight line (plane). (Notice that, intuitively, we may see a straight line (plane) as a circle (sphere) *with the center at infinity.*) In other words, a pencil of circles in a plane is the family of circles through a fixed, real or imaginary pair of points, and a pencil of spheres is the family of spheres through a fixed, real or imaginary circle. If the given circles (spheres, etc.) have the equations $\Gamma_1 = 0$, $\Gamma_2 = 0$, where the left hand sides are either of the form (3.1.3), (3.1.5) or linear (the case of a straight line or a plane), the equation of the corresponding pencil is

$$\lambda\Gamma_1 + \mu\Gamma_2 = 0, \tag{3.1.25}$$

where λ and μ are not simultaneously zero. Indeed, $\forall\lambda,\mu$, not both zero, (3.1.25) is the equation of a circle (sphere) or a line (plane) which passes through the intersection of Γ_1, Γ_2, and any other circle (sphere) of the pencil is defined if we give one more point P of it. But, if we ask P to satisfy (3.1.25), we get

$$\frac{\lambda}{\Gamma_2(P)} = \frac{\mu}{-\Gamma_1(P)}.$$

Hence, we get the values of λ, μ (up to a common factor, of course) for which the corresponding equation (3.1.24) represents the desired circle (sphere). Finally, like in the case of the pencil of straight lines (planes), we may divide the equation (3.1.24) by, say, λ and we see that the pencil consists of Γ_2, and of all the circles (spheres) of equation

$$\Gamma_1 + \nu\Gamma_2 = 0, \tag{3.1.25'}$$

for all the real values of the parameter ν. (Here, Γ_1 is supposed to be a circle (sphere).)

Of course, we may replace Γ_1, Γ_2 by any other pair of members of the pencil, and, in any case, one of the elements of the pencil is a real or imaginary straight line (plane) which is the common radical axis (plane) of all the circles (spheres) of the pencil.

We end this section by briefly considering some interesting surfaces which contain a whole family of circles. Let d be a straight line, and let

C be an arbitrary curve of space. If we *rotate* C around the *rotation axis* (*revolution axis*) d i.e., we let every point M of C move along a circle situated in the plane through M and perpendicular to d, with the center at the intersection of this plane with d, we obtain an envelope called a *surface of revolution* (see Fig. 3.1.9). The circles described by the various points of C are called *parallels* of the surface (and there is an infinite family of them, as promised). The curves obtained by the intersection of the surface by planes through the rotation axis are called *meridians*. Clearly, if we replace C by a meridian, the rotation (revolution) of the latter around d generates the same surface, and it is simpler to use a meridian in order to get the equation of the surface.

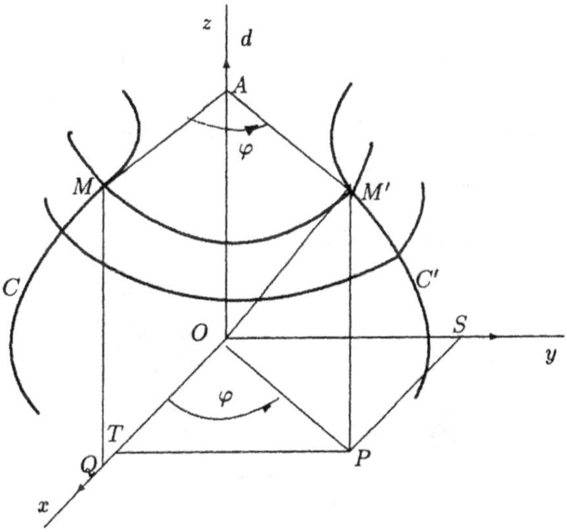

Fig. 3.1.9

Let us assume that C is a meridian, and let us take an orthogonal frame such that its z-axis is along d (with some arbitrary origin $O \in d$), and such that C is situated in the plane xOz (see again Fig. 3.1.9). Assume that C has the equations

$$F(x,z)=0,\ y=0. \tag{3.1.26}$$

This means that, $\forall M \in C$, and with the notation of Fig. 3.1.9, we have

$$0 = F(a(\vec{AM}), a(\vec{OA})) = F(a(\vec{AM'}), a(\vec{PM'}))$$
$$= F(\sqrt{x^2(M') + y^2(M')}, z(M')).$$

Since M' is just an arbitrary point of the surface, we may conclude that *the equation of the surface obtained by the revolution of C around the z-axis is*

(3.1.27) $$F(\sqrt{x^2 + y^2}, z) = 0,$$

i.e., *the equation is obtained by the simple replacement of x by $\sqrt{x^2 + y^2}$ in the equation of C in the plane (x, z).*

3.1.9 Examples. If C is $x^2 + z^2 - 1 = 0,\ y = 0$, the surface which we obtain is just the *unit sphere* (sphere of radius 1 and center at the origin) $x^2 + y^2 + z^2 - 1 = 0$. On the other hand, if C is a circle with the center on Ox, and not intersecting the z-axis:

$$(x - a)^2 + z^2 - b^2 = 0,\ y = 0, \quad (a > b > 0),$$

the corresponding surface (see Fig. 3.1.10) is called a *torus*, and it has the equation

(3.1.28) $$(\sqrt{x^2 + y^2} - a)^2 + z^2 - b^2 = 0.$$

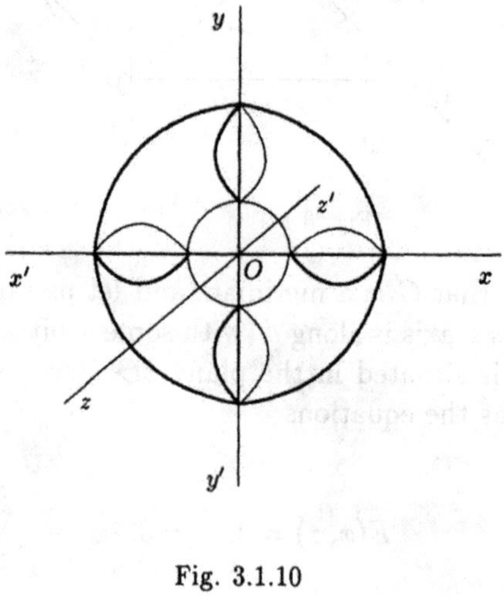

Fig. 3.1.10

We ask the reader to proceed in a similar way, and find the equation of a surface obtained by revolving a curve of the plane yOz around Oy, etc.

Finally, it is also easy to obtain parametric equations of, say, the surface obtained by rotating a curve C of the plane xOz around the z-axis. Indeed, assume that C has the parametric equations $x = f(t)$, $z = g(t)$, $y = 0$, and let us look at the generic point M' obtained by rotating M by the angle φ, as indicated in Fig.3.1.9. Then, we have

$$\begin{aligned} x(M') &= a(\vec{OT}) = a(\vec{OP})\cos\varphi = a(\vec{AM'})\cos\varphi \\ &= a(\vec{AM})\cos\varphi = f(t)\cos\varphi, \\ y(M') &= a(\vec{OS}) = a(\vec{OP})\sin\varphi = a(\vec{AM'})\sin\varphi \\ &= a(\vec{AM})\sin\varphi = f(t)\sin\varphi, \\ z(M') &= a(\vec{PM'}) = a(\vec{QM}) = g(t). \end{aligned}$$

Therefore, the parametric equations of the surface of revolution of C around Oz are

$$x = f(t)\cos\varphi,\ y = f(t)\sin\varphi,\ z = g(t). \tag{3.1.29}$$

As an exercise, we ask the reader to write down parametric equations of the torus (3.1.28).

Exercises and Problems.

3.1.1. Consider the circle of equation $(x-1)^2 + y^2 = 4$, and the point $P(2, -1/2)$, where x, y are orthogonal coordinates in a plane. i) Show that P belongs to the interior domain of the circle, and write down the equation of a straight line d which defines a chord of midpoint P. ii) Find the pole M of d with respect to the circle, and write down the quadratic equation of the pair of tangents to the circle through M.

3.1.2. Find the equations of the tangents through the origin to the circle

$$x^2 + y^2 - 10x - 4y + 25 = 0.$$

(Orthogonal coordinates.)

3.1.3. Find the angle between the circles

$$x^2 + y^2 = 16, \quad (x-5)^2 + y^2 = 9.$$

3.1.4. Show that the radical axis of two circles is perpendicular to the line which joins the centers of the circles and, also, that the radical axis is the locus of the centers of the circles which cut the given two circles under right angles.

3.1.5. Consider the circles

$$(\Gamma_1)\ (x-3)^2+y^2=24,\quad (\Gamma_2)\ x^2+(y+1)^2=16.$$

Find the equation of a third circle Γ which is orthogonal to Γ_1 and Γ_2, and passes through the point $(5,-4)$.

3.1.6. Let AB, CD be two orthogonal diameters of a given circle Γ. Let d be a variable straight line through C, and $M = d \cap line(AB)$, $N =$ the second intersection point of d and Γ. Show that the locus of the intersection point of the tangent to Γ at N with the parallel line to CD by M is the tangent of Γ at D.

3.1.7. The basis BC of a triangle is kept fixed and the third vertex A moves such that $\widehat{BAC} = constant$. Find the locus of the centroid (center of gravity) of ΔABC.

3.1.8. If Γ is a fixed circle, find the locus of the points M of the plane of Γ such that the tangents of Γ through M are orthogonal.

3.1.9. Let Γ be a circle of radius ρ, and $P \in \Gamma$, d a variable line through P, and $Q = \Gamma \cap d$. Find the locus of $M \in PQ$ such that $(P, M; Q) = \lambda = const.$

3.1.10. Write down the equation of a sphere which passes through the point $P(0,3,1)$, and through the circle $x^2+y^2=16,\ z=0$.

3.1.11. Compute the coordinates of the pole of the plane $ax+by+cz+d=0$ with respect to the sphere $x^2+y^2+z^2=\rho^2$. What condition must be satisfied by a,b,c,d for the plane to be tangent to the given sphere?

3.1.12. Consider the two spheres of equations

$$x^2+y^2+z^2=9,\quad x^2+y^2+z^2+4y-6z+7=0.$$

i) Compute the coordinates of the centers, and the radii of the given spheres. ii) Write down the equation of a sphere which contains the intersection circle of the given spheres, and the point $M(1,1,-1)$.

3.1.13. Find the tangent planes to the sphere

$$(x+5)^2 + (y-8)^2 + (z+1)^2 = 16,$$

which pass through the x-axis.

3.1.14. Find the equation of a sphere which is tangent to the lines

$$\frac{x-1}{3} = \frac{y+4}{6} = \frac{z-6}{4}, \quad \frac{x-4}{2} = \frac{y+3}{1} = \frac{z-2}{-6}$$

at the points $(1,-4,6)$, $(4,-3,2)$, respectively.

3.1.15. What is the locus of the points P which have power 7 with respect to the spheres

$$(x-5)^2 + (y-3)^2 + (z+1)^2 = 9,$$

$$(x-7)^2 + y^2 + (z-2)^2 = 16.$$

3.1.16. If Γ is a fixed sphere, and d is a fixed straight line, find the locus of the centers of the intersection circles of Γ with planes through d. Same question if the fixed line d is replaced by a fixed point A.

3.2 Conics and Quadrics

In this section, we will solve some classical problems which lead to curves and surfaces of quadratic equations more general than those of circles and spheres.

The first such problem is: *in plane, find the locus of the points M such that the sum of their distances to two given fixed points F_1, F_2 (called focus points) is a given, positive, real constant, say $2a$.* This locus is a curve called an *ellipse* and, with respect to a convenient frame called the *canonical frame* of the ellipse, it has a quadratic equation which we are going to find now. This equation of the ellipse will also be called *canonical*.

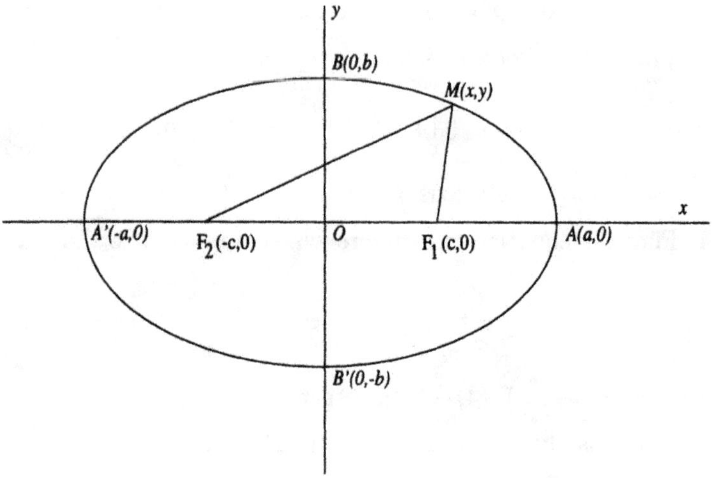

Fig. 3.2.1

Namely (see Fig. 3.2.1), we choose the midpoint O of the segment F_1F_2 as origin, the line F_1F_2 as the x-axis, and the perpendicular line to the latter at O as the y-axis. Then, if $l(F_1F_2) = 2c$, the characteristic condition for the points M of the ellipse $MF_1 + MF_2 = 2a$ becomes

$$\sqrt{(x-c)^2 + y^2} + \sqrt{(x+c)^2 + y^2} = 2a. \tag{3.2.1}$$

By moving the second square root to the right hand side, taking (twice) the square of the equation, and making reductions, equation (3.2.1) becomes

$$x^2(a^2 - c^2) + a^2y^2 = a^2(a^2 - c^2). \tag{3.2.2}$$

Finally, if we define $b = \sqrt{a^2 - c^2} > 0$, equation (3.2.2) may be put under the form

$$\frac{x^2}{a^2} + \frac{y^2}{b^2} - 1 = 0, \tag{3.2.3}$$

and this is the *canonical equation of the ellipse.* Any point $N(x_0, y_0)$ which satisfies (3.2.3) belongs to the ellipse. Indeed, if $NF_1 + NF_2 = 2a'$

we get as above

$$\frac{x^2}{a'^2} + \frac{y^2}{b'^2} - 1 = 0 \quad (b' = \sqrt{a'^2 - c^2}),$$

and a comparison with (3.2.3) yields

$$x_0^2\left(\frac{1}{a^2} - \frac{1}{a'^2}\right) + y_0^2\left(\frac{1}{b^2} - \frac{1}{b'^2}\right) = 0.$$

Since $a \le a'$, $a \ge a'$ imply $b \le b'$, $b \ge b'$, respectively, we get $a' = a$.

The numbers a, b are called *half-axes* of the ellipse. From (3.2.3), it is clear that the coordinate axes are symmetry axes of the curve (reader, please explain!), and the intersection points of these axes with the ellipse are called *vertices*. The origin is the *symmetry center* of the ellipse. The part of the ellipse which is situated in the first quadrant has the explicit equation

$$y = b\sqrt{1 - \frac{x^2}{a^2}}.$$

This shows that, for $0 \le x \le a$, y decreases from b to 0, and shows that the ellipse looks as indicated in Fig. 3.2.1. A circle may be seen as the particular case of an ellipse, where the two focuses coincide, and are situated at the center of the circle.

Similarly, we define a *hyperbola* as the locus of the points M such that $|MF_1 - MF_2| = 2a$. The notation is the same as above, and F_1, F_2 are the *focuses* of the hyperbola. The *canonical frame* of a hyperbola is defined in the same way as that of an ellipse, and the same computations as for an ellipse yield the *canonical equation* of the hyperbola

$$\frac{x^2}{a^2} - \frac{y^2}{b^2} - 1 = 0, \tag{3.2.4}$$

where $b = \sqrt{c^2 - a^2}$, and the fact that any point which satisfies (3.2.4) belongs to the hyperbola.

Again, the origin and the axes of the canonical frame are the symmetry center and axes of the curve, and the intersection points of these axes with the curve are *vertices*. (This time, the vertices on the y-axis are imaginary points.) In the first quadrant, the curve has the explicit equation

$$y = b\sqrt{\frac{x^2}{a^2} - 1},$$

and it is easy to understand that the curve looks as in Fig. 3.2.2. In particular, the curve consists of two branches. The two lines $y = \pm(b/a)x$ are the *asymptotes* of the hyperbola in the sense of the graph building methods of calculus: the points of the curve approach the asymptotes as much as we want as x goes to $\pm\infty$ (for $x \to \pm\infty$, $y/x \to \pm(b/a)$).

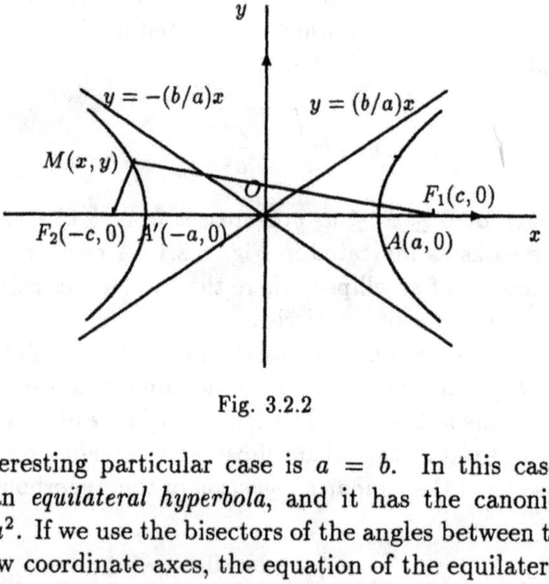

Fig. 3.2.2

An interesting particular case is $a = b$. In this case, the curve is called an *equilateral hyperbola*, and it has the canonical equation $x^2 - y^2 = a^2$. If we use the bisectors of the angles between the canonical axes as new coordinate axes, the equation of the equilateral hyperbola takes the following characteristic form $xy = k$, where k is a certain new constant. (Reader, please explain!)

Now, we will obtain a second definition of the ellipse and the hyperbola, and, also find a new interesting curve, the *parabola*.

In plane, let F be a fixed point, called a *focus*, and Δ be a fixed straight line which does not contain F, called a *directrix*. We will look for the locus of the points M of the plane which satisfy the condition

$$\frac{MF}{MN} = e = const. \qquad (MN \perp \Delta,\ N \in \Delta) \tag{3.2.5}$$

(see Fig. 3.2.3). The locus will be a curve, and the constant e is the *eccentricity* of the curve.

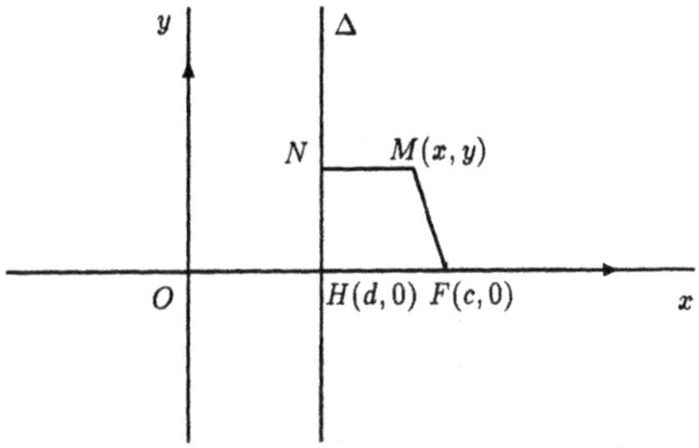

Fig. 3.2.3

In order to find this locus, we choose an orthogonal frame such that the x-axis is the perpendicular line from F to Δ (Fig. 3.2.3), we denote the coordinates of F by $(c, 0)$, and we suppose that the equation of Δ is $x = d$. (As the subsequent calculations show, it is not convenient to assume $d = 0$ here.)

Then, the characteristic condition (3.2.5) becomes

$$\sqrt{(x-c)^2 + y^2} = e|x - d|, \tag{3.2.6}$$

which, after a usual calculation, is equivalent to

$$x^2(1 - e^2) + y^2 + 2x(de^2 - c) + (c^2 - e^2d^2) = 0. \tag{3.2.7}$$

If $e \neq 1$, we may choose the y-axis such that $de^2 - c = 0$ i.e., $d/c = l(OH)/l(OF) = 1/e^2$ or, equivalently the simple ratio $(H, F; O) = 1/e^2$, a condition which obviously fixes the point O because H and F are given points. Then, the free term of (3.2.7) becomes

$$c^2 - e^2d^2 = d^2e^2(e^2 - 1),$$

and (3.2.7) is equivalent to

$$\frac{x^2}{d^2e^2} + \frac{y^2}{d^2e^2(1-e^2)} - 1 = 0. \tag{3.2.8}$$

Accordingly, if $e < 1$, the locus (3.2.8) is an ellipse of half-axes

$$a = ed = c/e, \qquad b = ed\sqrt{1-e^2}, \tag{3.2.9}$$

and, if $e > 1$, the locus is a hyperbola of half-axes

$$a = ed = c/e, \qquad b = ed\sqrt{e^2-1}. \tag{3.2.10}$$

Moreover, the converse results are also true i.e., for any ellipse (hyperbola) of canonical equation (3.2.3) ((3.2.4)), if we follow (3.2.9) ((3.2.10)) and define $e = c/a$, $d = a/e = a^2/c$, where c is the abscissa of the focus of the curve, the line $x = d$ plays the role of a directrix Δ, and the ellipse (hyperbola) is the locus of the points which satisfy (3.2.5). For symmetry reasons, it is clear that an ellipse (hyperbola) also has a second directrix $x = -d$ which together with the second focus $F_2(-c, 0)$ (Figs. 3.2.1, 3.2.2) satisfy again (3.2.5).

Now, let us come back to equation (3.2.7), and discuss the locus in the case $e = 1$. In this case, the locus is called a *parabola*. In other words, a *parabola* is a locus of points which are at equal distance from a fixed point called *focus* and a fixed straight line called *directrix*. Before any special choice of the y-axis, the equation of the parabola is

$$y^2 + 2x(d-c) + (c^2 - d^2) = 0. \tag{3.2.11}$$

This shows that it is convenient to define a *canonical frame* of the parabola by taking the origin at the midpoint of the segment HF (see Fig. 3.2.4, i.e., such that $d = -c$. With respect to this frame, equation (3.2.11) becomes the *canonical equation of the parabola*

$$y^2 - 2px = 0, \tag{3.2.12}$$

where $p := 2c$, and p is called the *parameter* of the parabola.

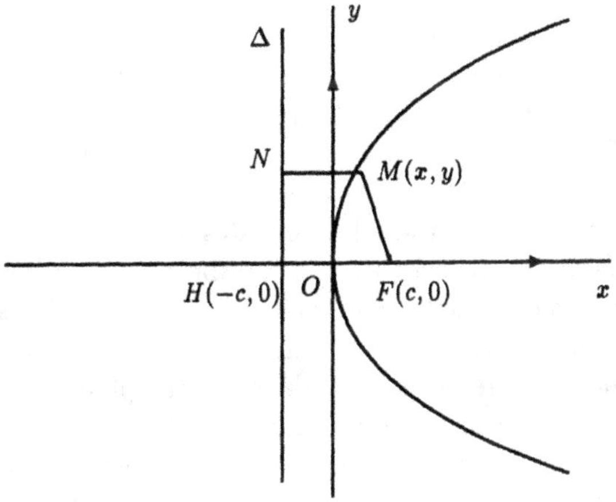

Fig. 3.2.4

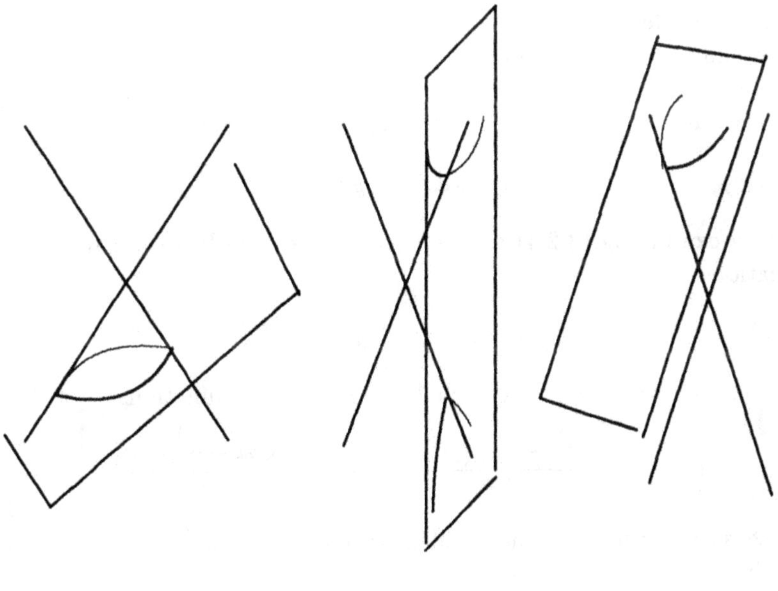

Fig. 3.2.5

It is clear that the parabola passes through O, called the *vertex* of the parabola, and that the x-axis is a symmetry axis of the parabola; we only have one symmetry axis this time. It is easy to understand that the geometric form of the parabola is the one shown in Fig. 3.2.4. (Remember the discussion of the graph of a quadratic function in high school.)

There exists a term which designates all the curves discussed above namely, ellipses, hyperbolas and parabolas are called *nondegenerate conic sections* or, briefly, *conics.* The reason for this name is that one can prove that all these curves appear as sections of the infinite surface of a *revolution cone* by various planes: if the plane cuts only one sheet of the cone we get an ellipse, if it cuts both we get a hyperbola, and if the plane is parallel with a generator of the cone we get a parabola (see Fig. 3.2.5). The qualification *nondegenerate* was added to the name of the curves to distinguish them from the sections of the cone by planes that pass through the vertex of the cone; such sections are just pairs of straight lines, and we may refer to such pairs as *degenerate conics.* We do not include here the proof of the above results.

Finally, let us also mention that there are simple parametric equations of the ellipses, hyperbolas and parabolas. Namely, by examining (3.2.3), we see that the points of an ellipse may be defined by

$$x = a\cos t, \qquad y = b\sin t, \tag{3.2.13}$$

and by examining (3.2.4) we see that a hyperbola has the parametric equations

$$x = a\cosh t, \qquad y = b\sinh t. \tag{3.2.14}$$

(In (3.2.14), we have the hyperbolic trigonometric functions

$$\sinh t := \frac{\exp t - \exp(-t)}{2}, \ \cosh t := \frac{\exp t + \exp(-t)}{2},$$

which satisfy the well known identity $\cosh^2 t - \sinh^2 t = 1$.) For a parabola, we may just write

$$x = t^2/2p, \quad y = t \tag{3.2.15}$$

instead of (3.2.12).

In space, the surfaces which have an implicit equation of the second degree are designated by the natural name of *quadrics*. In the remainder of this section, we will use the conics in order to generate examples of quadrics.

If the ellipse (3.2.3) is rotated around the x-axis, a revolution surface called an *ellipsoid of revolution* is obtained. The reader can easily imagine the geometric form of the surface. The spatial orthogonal frame which consists of the canonical frame of the ellipse and the perpendicular to the plane of the ellipse at its center as z-axis is called the *canonical frame* of the ellipsoid. Following the results of section 3.1 concerning revolution surfaces, we see that the equation of our revolution ellipsoid is

$$\frac{x^2}{a^2} + \frac{y^2 + z^2}{b^2} - 1 = 0. \tag{3.2.16}$$

The obtained result suggests us to define a *general ellipsoid* as a surface which has the equation

$$\frac{x^2}{a^2} + \frac{y^2}{b^2} + \frac{z^2}{c^2} - 1 = 0 \tag{3.2.17}$$

with respect to a certain orthogonal frame called the *canonical frame* of the ellipsoid. It is clear that the ellipsoid is symmetric with respect to the origin (*the center of the ellipsoid*), with respect to the coordinate planes (*the symmetry planes*) and with respect to the canonical coordinate axes (*the axes of the ellipsoid*). The numbers a, b, c are called the *half-axes* of the ellipsoid. The sections of the ellipsoid (3.2.17) by planes which are parallel to the coordinate planes clearly are ellipses, and this shows that an ellipsoid is a surface which looks like the envelope of the body represented in Fig. 3.2.6.

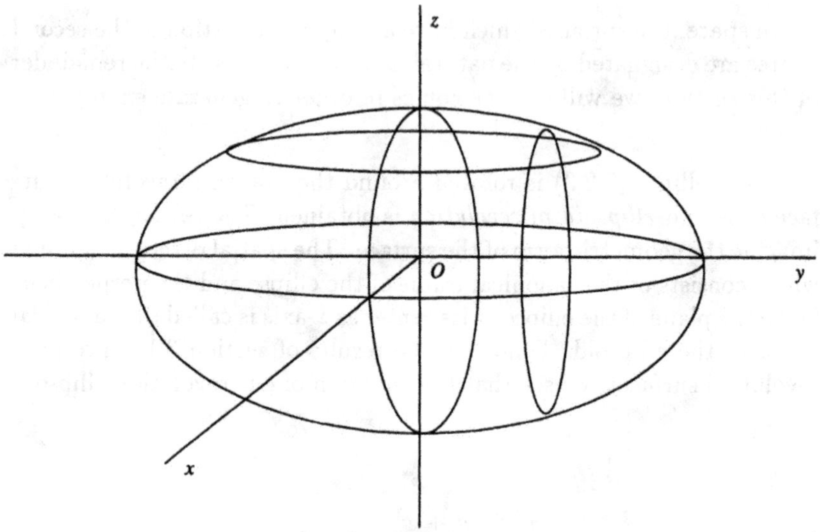

Fig. 3.2.6

Now, let us proceed in a similar way with the hyperbola (3.2.4). First, we rotate it around the x-axis, and get the revolution surface

$$\frac{x^2}{a^2} - \frac{y^2 + z^2}{b^2} - 1 = 0. \tag{3.2.18}$$

The geometric form of this surface is seen by rotating Fig. 3.2.2 around the x-axis; the surface has two sheets, and we will call it a *two-sheeted revolution hyperboloid.* Furthermore, in analogy with equation (3.2.18), we write down the equation

$$\frac{x^2}{a^2} - \frac{y^2}{b^2} - \frac{z^2}{c^2} - 1 = 0, \tag{3.2.19}$$

and we call the surface represented by this equation a *general two-sheeted hyperboloid*, while the frame with respect to which the surface has equation (3.2.19) is the *canonical frame.* The surface has a center, planes and axes of symmetry just like the ellipsoid. Its intersections with planes $x = const.$ are ellipses and the intersection with planes $y = const.$ or $z = const.$ are hyperbolas. Accordingly, the surface looks as indicated in Fig. 3.2.7. In Fig. 3.2.7, we also see the cone with vertex at origin, and of equation

$$\frac{x^2}{a^2} - \frac{y^2}{b^2} - \frac{z^2}{c^2} = 0, \tag{3.2.20}$$

which is defined by the rotation of the asymptotes of the hyperbola in the case of a revolution hyperboloid. The cone (3.2.20) is called the *asymptotic cone* of the hyperboloid.

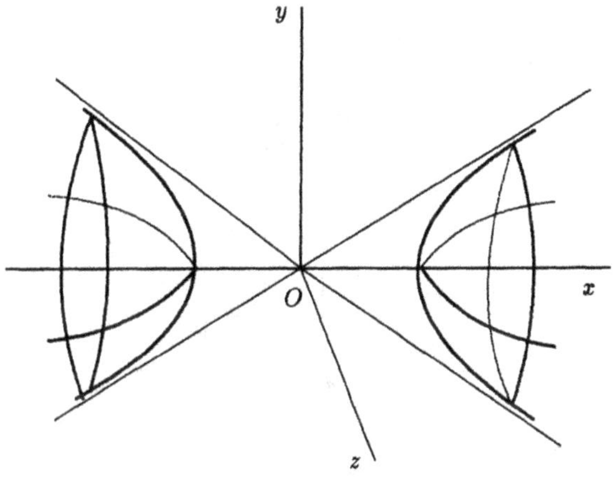

Fig. 3.2.7

On the other hand, if we rotate the hyperbola (3.2.4) around the y-axis, we get a one-sheeted surface, called a *one-sheeted revolution hyperboloid*, which has the equation

$$\frac{x^2 + z^2}{a^2} - \frac{y^2}{b^2} - 1 = 0. \tag{3.2.21}$$

The corresponding *general one-sheeted hyperboloid* will be defined as the surface of *canonical equation*

$$\frac{x^2}{a^2} + \frac{y^2}{b^2} - \frac{z^2}{c^2} - 1 = 0. \tag{3.2.22}$$

(Notice that we changed the role of the coordinates y and z, while writing this equation.) This surface again has a center, axes and planes of symmetry. The planes $x = const$ and $y = const.$ cut the surface by hyperbolas, and the planes $z = const.$ cut it by ellipses. The surface looks as the one indicated in Fig. 3.2.8. It has the characteristic form of a saddle, and it has the asymptotic cone

$$\frac{x^2}{a^2} + \frac{y^2}{b^2} - \frac{z^2}{c^2} = 0. \tag{3.2.23}$$

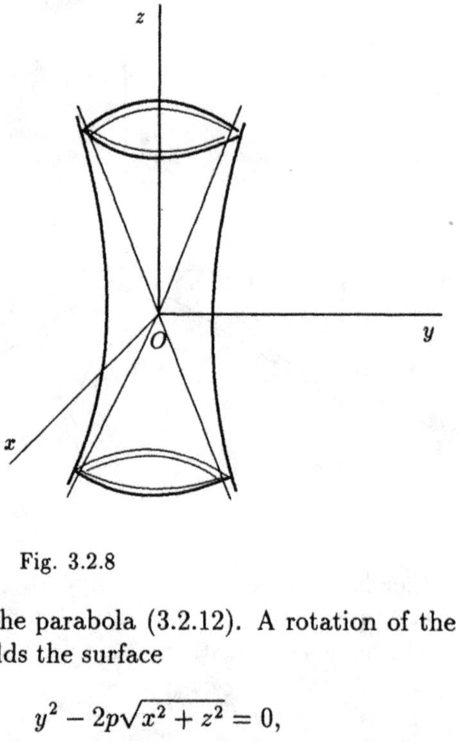

Fig. 3.2.8

Now, we look at the parabola (3.2.12). A rotation of the parabola around the y-axis yields the surface

$$y^2 - 2p\sqrt{x^2 + z^2} = 0,$$

which is not of the second degree hence, it is not a quadric, and we have no reason to discuss it here. But, the rotation of the parabola around its symmetry axis x, yields the quadric

$$y^2 + z^2 - 2px = 0 \tag{3.2.24}$$

naturally called a *paraboloid of revolution*.

The equation (3.2.24) suggests us to consider quadrics of canonical equations

(3.2.25) $$\frac{x^2}{a^2} \pm \frac{y^2}{b^2} - 2z = 0.$$

In the case of the sign +, the surface is called an *elliptic paraboloid.* It is symmetric with respect to the planes $x = 0$, $y = 0$ (but not $z = 0$), and with respect to the z-axis. It is cut by planes $z = const.$ following ellipses and by planes $x = const.$, $y = const.$ following parabolas. The surface looks as indicated by Fig. 3.2.9.

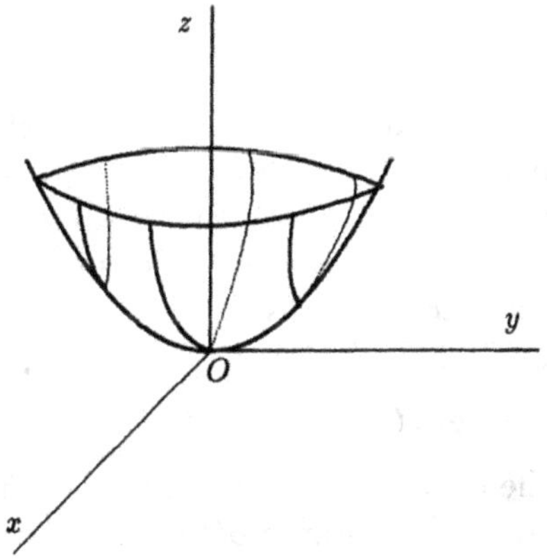

Fig. 3.2.9

In the case of equation (3.2.25) with the sign minus, the surface is called a *hyperbolic paraboloid.* It has the same symmetries as the elliptic paraboloids and, again, the intersections with planes $x = const.$, $y = const.$ are parabolas but, the intersection with planes $z = const.$ are hyperbolas. The surface looks as indicated by Fig. 3.2.10.

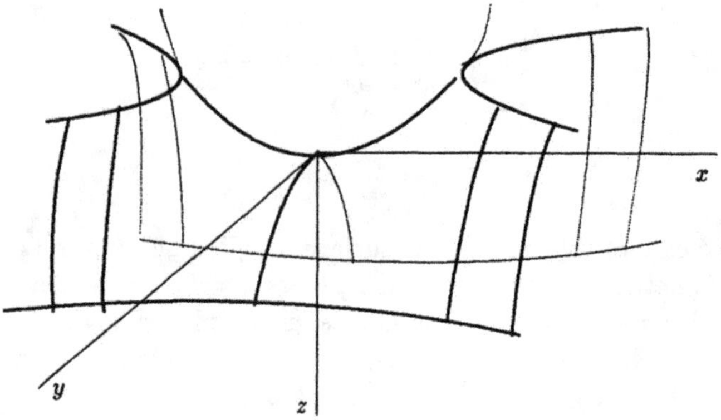

Fig. 3.2.10

The quadrics defined above i.e., ellipsoids, hyperboloids and paraboloids are called *nondegenerate quadrics*. The other types of possible quadrics (e.g., a pair of planes with the quadratic equation given by the multiplication of the two linear equations of the two planes; there are also more complicated examples, quadratic cylinders and cones) are called *degenerate quadrics*.

Exercises and Problems

3.2.1. Show that the equation $25x^2 + 169y^2 = 4225$ defines an ellipse. Compute the half axes, the distance between its focuses, and the equation of its directrices.

3.2.2. Find two points P, Q of the ellipse $x^2/36 + y^2/9 = 1$ which together with $A(6,0)$ form an equilateral triangle.

3.2.3. Prove that if an equilateral hyperbola contains the vertices of a triangle ΔABC it also contains the orthocenter of the triangle.

3.2.4. Let Δ be a rectangle with the vertices on the ellipse $x^2/49 + y^2/24 = 1$, and with two opposite edges perpendicular to Ox through the focal points. Compute the area of Δ.

3.2.5. The endpoints of a straight line segment of a fixed length l move along a given pair of orthogonal axes. Find the locus of $M \in line(AB)$, if $(A, B; M) = k = const.$

3.2.6. Find the focal points, the eccentricity, the directrices and the asymptotes of the hyperbola $x^2/9 - y^2/16 = 1$.

3.2.7. Find the canonical equation of a hyperbola of asymptotes $y = \pm(1/2)x$ which passes through the point $(12, \sqrt{3})$.

3.2.8. For a given hyperbola, let F_1 be a focus, d_1 the corresponding directrix and α an asymptote. If $F_1P \perp \alpha$, $P \in \alpha$, show that $P \in d_1$.

3.2.9. Let P move along the circle $x^2 + y^2 = r^2$, and let M be the point which divides the ordinate of P into segments of ratio λ. Find the locus of M.

3.2.10. Find the locus of the centers of the circles which are tangent to two given circles.

3.2.11. Write the canonical equation of the hyperbolas for which $P(16/5, 12/5)$ is the intersection point of an asymptote with a directrix.

3.2.12. Find a square with the edges parallel to the coordinate axes, and the vertices on the hyperbola $x^2/a^2 - y^2/b^2 = 1$. When does a solution exist?

3.2.13. The triangle ΔABC has the vertex A at the vertex of the parabola $y^2 - 8x = 0$, and its orthocenter M at the focus of the same parabola. Find the equations of the sides of ΔABC.

3.2.14. Compute the length of the sides of an equilateral triangle whose vertices are on the parabola $y^2 - 2ox = 0$.

3.2.15. Show that all the ellipses obtained by the intersection of the ellipsoid

$$\frac{x^2}{a^2} + \frac{y^2}{b^2} + \frac{z^2}{c^2} - 1 = 0$$

by planes $x = k = const.$ have the same eccentricity.

3.2.16. Show that the plane $2x + 3y - 6z - 6 = 0$ cuts the hyperboloid

$$\frac{x^2}{9} + \frac{y^2}{4} - z^2 - 1 = 0$$

by two straight lines.

3.2.17. Find the locus of the points situated on straight lines which meet the lines

$$\frac{x}{2} = \frac{y-1}{0} = \frac{z}{-1}, \quad \frac{x-2}{0} = \frac{y}{1} = \frac{z}{1}, \quad \frac{x}{2} = \frac{y+1}{0} = \frac{z}{1}.$$

3.2.18. Find the focus of the parabola of intersection of the paraboloid $x^2/16 - y^2/4 = z$ by the plane $y = 2$.

3.2.19. In space, find the locus of the points M such that $MA + MA' =$ *const.* ($MA - MA' =$ *const.*), where A, A' are two fixed points.

3.3 Conics and Quadrics: general theory

By definition, quadratic geometry studies all the geometric figures in plane and space which are represented by a quadratic equation with respect to an affine frame, and, as already said, the plane figures of this type are called *conics*, while those of space are called *quadrics*.

It is important to notice that *the above definitions are invariant with respect to affine changes of the frame.* Indeed, such changes are given by linear expressions of the type (1.3.10) hence, the change transforms a quadratic equation into a quadratic equation.

The most general plane quadratic equation is of the following form:

(3.3.1) $$a_{11}x^2 + a_{22}y^2 + 2a_{12}xy + 2a_{10}x + 2a_{20}y + a_{00} = 0,$$

where the indices of the coefficients a_{ij} $(i, j = 0, 1, 2)$ were chosen such that the index 1 corresponds to the coordinate x, 2 corresponds to y and 0 to a (nonwritten) factor 1 in the corresponding term of (3.3.1). In space, the similar equation is:

(3.3.2) $$a_{11}x^2 + a_{22}y^2 + a_{33}z^2 + 2a_{12}xy + 2a_{13}xz + \\ +2a_{23}yz + 2a_{10}x + 2a_{20}y + 2a_{30}z + a_{00} = 0,$$

where the new index 3 corresponds to z.

In order to make general computations easy, we will write the equations (3.3.1) and (3.3.2) by using vectors and matrices. We give the necessary explanations for the space equation, and we ask the reader to particularize to plane by himself. We begin by the case of orthogonal coordinates. Then, we may see (x, y, z) as the coordinates of the radius vector $\bar{r}$, and we define a *transformation* or *operator* T as follows: with the vector $\bar{r}$, we associate a new vector denoted $T\bar{r}$ which has the coordinates

$$T\bar{r}\ (a_{11}x + a_{12}y + a_{13}z,\ \ a_{21}x + a_{22}y + a_{23}z,\ \ a_{31}x + a_{32}y + a_{33}z), \tag{3.3.3}$$

where the coefficients that did not appear in (3.3.2) are defined by the *symmetry condition* $a_{ij} = a_{ji}$. Furthermore, we introduce the vector

$$\bar{a}(a_{10}, a_{20}, a_{30}) \tag{3.3.4}$$

and denote

$$\alpha := a_{00}. \tag{3.3.5}$$

Then, equation (3.3.2) becomes

$$f(\bar{r}) := \bar{r}.T\bar{r} + 2\bar{a}.\bar{r} + \alpha = 0. \tag{3.3.6}$$

If we consider 2-dimensional vectors (i.e., $z = 0$, and all the coefficients with an index 3 zero), the same equation (3.3.6) is the equation of the general conic (3.3.1).

3.3.1 Lemma. *The operator T has the following properties*

$$T(\alpha\bar{u} + \beta\bar{v}) = \alpha T\bar{u} + \beta T\bar{v},$$

$$\bar{u}.T\bar{v} = \bar{v}.T\bar{u}.$$

Proof. Of course, in order to apply T to another vector we will use the coordinates of this new vector instead of (x, y, z) of (3.3.3). The announced results follow by straightforward calculations. The first property of the Lemma is known as the *linearity* of the *operator* T, and the second property is known as its *symmetry*, and is a result of the symmetry condition imposed on the coefficients a_{ij}. Q.E.D.

Now, let us consider the case of general affine coordinates. With a conic, we may associate the following matrices

$$(3.3.7) \qquad A := \begin{pmatrix} a_{11} & a_{12} \\ a_{21} & a_{22} \end{pmatrix}, \quad \tilde{A} := \begin{pmatrix} a_{11} & a_{12} & a_{10} \\ a_{21} & a_{22} & a_{20} \\ a_{01} & a_{02} & a_{00} \end{pmatrix}.$$

Similarly, in the case of a quadric we define

$$(3.3.8) \qquad A := \begin{pmatrix} a_{11} & a_{12} & a_{13} \\ a_{21} & a_{22} & a_{23} \\ a_{31} & a_{32} & a_{33} \end{pmatrix}, \quad \tilde{A} := \begin{pmatrix} a_{11} & a_{12} & a_{13} & a_{10} \\ a_{21} & a_{22} & a_{23} & a_{20} \\ a_{31} & a_{32} & a_{33} & a_{30} \\ a_{01} & a_{02} & a_{03} & a_{00} \end{pmatrix}.$$

We will refer to A as the *small matrix*, and to $\tilde{A}$ as the *large matrix* of the conic (quadric).

Furthermore, we interpret the coordinates of a vector as a one-column matrix, and define

$$(3.3.9) \qquad \xi := \begin{pmatrix} x \\ y \\ z \end{pmatrix}, \quad \mathcal{X} := \begin{pmatrix} x \\ y \\ z \\ 1 \end{pmatrix}, \quad a := \begin{pmatrix} a_{10} \\ a_{20} \\ a_{30} \end{pmatrix}.$$

We will use the same notation for a conic, but the columns (3.3.9) will have one element less. Then, equation (3.3.6) takes the form

$$(3.3.10) \qquad \xi^t A\xi + 2a^t\xi + \alpha = 0,$$

and also

$$(3.3.11) \qquad \mathcal{X}^t \tilde{A} \mathcal{X} = 0.$$

$A\xi$ is the application of the operator T defined as in the orthogonal case, and A is the matrix of the operator T. The upper index t denotes the transposed matrix. Equation (3.3.10) may be seen as being obtained by the same calculations as (3.3.6) while the possibly nonorthogonal character of the coordinate system is "forgotten".

Since we want to emphasize geometry, our approach will be to use equation (3.3.6), and work with vectors, even in affine coordinates, just as if the coordinates were orthogonal coordinates. The mathematical justification of this approach is matrix algebra. We will just have to keep in mind that, in (3.3.6), the dots are not usual scalar products unless in orthogonal coordinates. If the problems and results discussed have an affine character, these results are valid in general affine coordinates. But, if notions such as distance, angle, orthogonality come into play, the results will only be valid in orthogonal coordinates. Most of the following results are affine. Orthogonal coordinates will be necessary in our future discussion of symmetry axes and planes of conics and quadrics.

The following Proposition yields some simple interesting facts which are consequences of the general equations (3.3.1) and (3.3.2).

3.3.2 Proposition. *Generically, there exists one conic which passes through five given points, and one quadric which passes through nine given points. The intersection of a quadric with a plane is a conic.*

Proof. If five points are asked to satisfy equation (3.3.1), one gets a system of five linear *homogeneous* equations with six unknowns a_{ij}. Generically (i.e., if the equations are independent), such a system determines a_{ij} up to a common factor, and this proves our assertion about conics. Similarly, for a quadric, we have nine equations with ten unknowns. For the last assertion of the proposition, if we use equation (3.3.2) of the quadric, and parametric equations (2.2.15) of the plane, the intersection has a quadratic equation with respect to interior coordinates in the plane. Q.E.D.

Now, we are going to develop some of the geometry of the general quadratic equations. Most of it will come from the discussion of the intersection of a conic or quadric with a straight line.

3.3.3 Proposition. *A generic straight line d intersects a conic (quadric) Γ at two points which may be real and different, real and coincident or imaginary points.*

Proof. We assume that Γ has equation (3.3.6), and d is represented by

$$\bar{r} = \bar{r}_0 + \lambda \bar{v}, \tag{3.3.12}$$

where $\bar{r}_0$ is the radius vector of a point M_0 of d, and $\bar{v}$ defines the direction of d. If (3.3.12) is inserted into (3.3.6), and if we use the properties of the operator T as given in Lemma 3.3.1, the resulting equation for λ may be written as

$$\lambda^2(\bar{v}.T\bar{v}) + 2\lambda B(\bar{r}_0, \bar{v}) + f(\bar{r}_0) = 0, \tag{3.3.13}$$

where

$$B(\bar{r}_0, \bar{v}) := \bar{v}.T\bar{r}_0 + \bar{v}.\bar{a} = \bar{r}_0.T\bar{v} + \bar{a}.\bar{v}, \tag{3.3.14}$$

and $f(\bar{r}_0)$ is the left hand side of (3.3.6) evaluated on $\bar{r}_0$ instead of $\bar{r}$.

Since $d \cap \Gamma$ consists of the points of radius vector (3.3.12) where λ are solutions of (3.3.13), we are done. Q.E.D.

We will refer to equation (3.3.13) as the *intersection equation.*

Before anything else, let us examine the nongeneric cases.

3.3.4 Definition. *If the intersection equation (3.3.13) of a certain line d is linear, rather than quadratic, i.e., if the direction of d satisfies the condition*

$$\bar{v}.T\bar{v} = 0, \tag{3.3.15}$$

the direction of d is called an asymptotic direction of the conic (quadric).

Of course, *the direction of a straight line d is asymptotic iff d intersects the conic (quadric) at no more than one point.* If the vector $\bar{v}$ defines an asymptotic direction, we will also say that $\bar{v}$ is an *asymptotic vector.*

Since the left hand side of equation (3.3.15) is homogeneous of the second degree with respect to the coordinates of the vector $\bar{v}$, it is clear that, in space, the straight lines of asymptotic direction and passing through any fixed point A generate a cone, called the *cone of asymptotic directions* of the quadric at A. Similarly, in plane, we get a *pair of asymptotic directions* of a conic. Notice that *equation (3.3.15) of the asymptotic directions at the origin is given by equating to zero the group of terms of second degree of the equation of the conic (quadric).* The cone or the pair of asymptotic directions may be imaginary.

3.3.5 Definition. *If for a certain line d the intersection equation (3.3.13) has no solutions, which happens iff*

$$(3.3.16) \qquad \bar{v}.T\bar{v} = 0, \quad \bar{r}_0.T\bar{v} + \bar{a}.\bar{v} = 0, \quad f(\bar{r}_0) \neq 0,$$

the line d is called an asymptote of the conic (quadric).

This definition means that d is an asymptote iff $d \cap \Gamma$ is *void* in the sense that it contains no points at all, neither real nor imaginary.

Let us concentrate on a conic first. Since the point M_0 of d may be arbitrary, it follows from (3.3.16) that the asymptotes of the conic are precisely the straight lines d of equation

$$(3.3.17) \qquad \bar{r}.T\bar{v} + \bar{a}.\bar{v} = 0,$$

where $\bar{v}$ is an asymptotic vector for which $T\bar{v} \neq 0$, and under the restriction that the line does not cut the conic.

In the case of a quadric, the asymptotes are again characterized by (3.3.16), and (3.3.17) must hold *along an asymptote*. But, now, (3.3.17) defines a plane associated with the asymptotic direction $\bar{v}$. Such a plane is called an *asymptotic plane* of the quadric. The intersection of the quadric with an asymptotic plane is a conic (see Proposition 3.3.2), and it is clear that an asymptote d of direction $\bar{v}$ which sits in the given asymptotic plane (if d exists) is also an asymptote of the intersection conic, and conversely. (Indeed, d cannot intersect this conic because the intersection point would be an intersection point of d with the quadric. Conversely, if d does not intersect the conic, it will not intersect the quadric either.) Therefore, in order to find all the asymptotes of a quadric, we will find all the asymptotic planes, and then, the asymptotes of their intersection conics with the quadrics (using interior coordinates, for instance).

3.3.6 Proposition. *For the nondegenerate conics defined by the canonical equations of Section 3.2, there are: 2 real asymptotes for a hyperbola, 2 imaginary asymptotes for an ellipse, one asymptotic direction, but no asymptotes for a parabola.*

Proof. The asymptotic directions at the origin are given by:

$$\frac{x^2}{a^2} - \frac{y^2}{b^2} = 0 \quad \text{(hyperbola)},$$

$$\frac{x^2}{a^2} + \frac{y^2}{b^2} = 0 \quad \text{(ellipse)},$$
$$y^2 = 0 \qquad \text{(parabola)}.$$

Furthermore, for a hyperbola the equation (3.3.17) associated with the two real asymptotic directions obtained above yields the asymptotes

$$\frac{x}{a} \pm \frac{y}{b} = 0,$$

i.e., we find again the same asymptotes as in Section 3.2.

For an ellipse (3.3.17) is

$$\frac{x}{a} \pm \sqrt{-1}\frac{y}{b} = 0.$$

For a parabola, (3.3.17) would be $p = 0$ (the asymptotic direction satisfies $T\bar{v} = 0$), which doesn't hold, and there is no asymptote. Q.E.D.

A similar analysis can be made for the nondegenerate quadrics with the canonical equations of Section 3.2, but it is more complicated. The equations of the asymptotic cones at the origin are:

$$\frac{x^2}{a^2} \pm \frac{y^2}{b^2} - \frac{z^2}{c^2} = 0 \quad \text{(hyperboloids)},$$

$$\frac{x^2}{a^2} + \frac{y^2}{b^2} + \frac{z^2}{c^2} = 0 \quad \text{(ellipsoids)},$$

$$\frac{x^2}{a^2} \pm \frac{y^2}{b^2} = 0 \quad \text{(paraboloids)}.$$

Thus, for hyperboloids this cone is real, and, by comparing with the equation of the hyperboloid, we see immediately that all the generators of the cone are asymptotes. We call this cone with the vertex at the canonical origin (the center) the *asymptotic cone of the hyperboloid.* For the ellipsoids, the asymptotic cone is an imaginary cone. For the paraboloids, the cone reduces to a pair of intersecting plane which are real for the hyperbolic paraboloid and imaginary for the elliptic paraboloid. But, we can find more real asymptotes for hyperboloids. For instance, take any plane which cuts the hyperboloid by a hyperbola; then, the asymptotes of this hyperbola will be asymptotes of the hyperboloid as well. The same can be done in the case of a hyperbolic paraboloid.

The next possible nongeneric case of an intersection between a conic (quadric) and a straight line is:

3.3.7 Definition. *The straight line d may have all its points on the conic (quadric), and, then, d is called a rectilinear generator of the conic (quadric).*

This notion is not interesting for conics since it is clear that we find rectilinear generators only if the conic *degenerates* into a pair of straight lines. But, we can find interesting rectilinear generators for some nondegenerate quadrics. Of course, d is a rectilinear generator if all the coefficients of the intersection equation (3.3.13) vanish i.e., the equality to zero holds for any λ. We will not make a general discussion of this subject, and content ourselves only with:

3.3.8 Proposition. *The one-sheeted hyperboloids and the hyperbolic paraboloids have two families of real rectilinear generators.*

Proof. A straightforward inspection of the equation (3.2.22) of a one-sheeted hyperboloid shows that, for any value of the parameter $t \neq 0$, the lines

$$\frac{x}{a} - \frac{z}{c} = t\left(1 - \frac{y}{b}\right), \quad \frac{x}{a} + \frac{z}{c} = \frac{1}{t}\left(1 + \frac{y}{b}\right), \tag{3.3.18}$$

$$\frac{x}{a} - \frac{z}{c} = t\left(1 + \frac{y}{b}\right), \quad \frac{x}{a} + \frac{z}{c} = \frac{1}{t}\left(1 - \frac{y}{b}\right) \tag{3.3.19}$$

are situated on the hyperboloid. These are the two families of rectilinear generators announced by Proposition 3.3.8.

Similarly, from the canonical equation (3.2.25), we get the rectilinear generators of the hyperbolic paraboloid:

$$\frac{x}{a} - \frac{y}{b} = 2t, \quad \frac{x}{a} + \frac{y}{b} = \frac{z}{t}, \tag{3.3.20}$$

$$\frac{x}{a} + \frac{y}{b} = 2t, \quad \frac{x}{a} - \frac{y}{b} = \frac{z}{t}. \tag{3.3.21}$$

Q.E.D.

One can see that the other nondegenerate quadrics do not have real rectilinear generators.

Now, we return to the general *quadratic* intersection equation (3.3.13).

3.3.9 Definition. *If a straight line d intersects a conic (quadric) Γ at two equal points $M_1 = M_2 = M$ (i.e., at one point counted twice), the line is said to be tangent to the conic (quadric) at the contact point M.*

From (3.3.13), we see that, if d and Γ are defined by the equations indicated in the proof of Proposition 3.3.3, d is a tangent to Γ iff

$$B^2(\bar{r}_0, \bar{v}) - (\bar{v}.T\bar{v})f(\bar{r}_0) = 0. \tag{3.3.22}$$

We call this condition the *tangency equation.*

First, we would like to determine all the tangents of Γ through a given fixed point M_0 of radius vector $\bar{r}_0$. If M of radius vector $\bar{r}$ belongs to such a tangent d, the direction of d may be taken as that of $\bar{v} = \bar{r} - \bar{r}_0$ (see the earlier Fig. 3.1.6, where Γ is the circle). Correspondingly, all the required tangents are represented by the tangency equation where $\bar{v}$ is replaced by $\bar{r} - \bar{r}_0$ i.e.,

$$B^2(\bar{r}_0, \bar{r} - \bar{r}_0) - [(\bar{r} - \bar{r}_0).T(\bar{r} - \bar{r}_0)]f(\bar{r}_0) = 0. \tag{3.3.23}$$

Now, if we use (3.3.14), by an easy calculation, we see that

$$B(\bar{r}_0, \bar{r} - \bar{r}_0) = f(\bar{r}; \bar{r}_0) - f(\bar{r}_0), \tag{3.3.24}$$

where

$$f(\bar{r}; \bar{r}_0) := \bar{r}.T\bar{r}_0 + \bar{a}.(\bar{r} + \bar{r}_0) + \alpha, \tag{3.3.25}$$

the notation being that of equation (3.3.6) of Γ.

The expression $f(\bar{r}; \bar{r}_0)$ is called the *polarized form* of the equation of the conic (quadric) with respect to the point M_0. This polarization operation of an equation of the second degree is an extension of the polarization operation (3.1.21) encountered in the case of a circle (sphere). If the vectors are expressed by the corresponding coordinates, the *polarization is obtained by the rules (3.1.21) supplemented by the replacements of products as follows:*

$$xy \longmapsto \frac{1}{2}(xy_0 + yx_0). \tag{3.3.26}$$

Now, with (3.3.24) and (3.3.25), equation (3.3.23) becomes (after some algebraic computations)

$$f^2(\bar{r}; \bar{r}_0) - f(\bar{r})f(\bar{r}_0) = 0. \tag{3.3.27}$$

Consequently, we get

3.3.10 Proposition. *If Γ is a conic, and M_0 is not on Γ, there are two, possibly imaginary, tangents of Γ through M_0. Iff $M_0 \in \Gamma$, these two tangents coincide, the corresponding line is the tangent of Γ at M_0, and it has the equation*

$$f(\bar{r}; \bar{r}_0) = 0 \tag{3.3.28}$$

obtained by polarization. With the similar notation, in the case of a quadric, if M_0 is not on Γ, there exists a (possibly imaginary) quadratic (i.e., of quadratic equation) cone of tangents of Γ with vertex M_0, and, if $M_0 \in \Gamma$, this cone reduces to a (double) plane, called the tangent plane of the quadric at M_0. The equation of the tangent plane is also given by polarization i.e., the equation is (3.3.28).

Proof. That we have two tangents in the case of a conic or a quadratic cone in the case of a quadric follows from the fact that equation (3.3.27) is of second degree. Furthermore, if $M_0 \in \Gamma$, we have $f(\bar{r}_0) = 0$, and the equation (3.3.27) reduces to (3.3.28). Q.E.D.

3.3.11 Example. Let Γ be the quadric of equation

$$x^2 + 2y^2 - 3z^2 + xy + 3xz - 2yz + 2x - y + z - 4 = 0, \tag{3.3.29}$$

and $M_0(1,1,1)$. Check that $M_0 \in \Gamma$, and write down the equation of the tangent plane of Γ at M_0.

Solution. To check that the coordinates of M_0 satisfy (3.3.29) is straightforward. The tangent plane is obtained by polarization:

$$1.x + 2.1.y - 3.1.z + \frac{1}{2}(x.1 + 1.y) + \frac{3}{2}(x.1 + 1.z) - \frac{2}{2}(y.1 + 1.z) + \frac{2}{2}(x+1) -$$

$$-\frac{1}{2}(y+1) + \frac{1}{2}(z+1) - 4 = 0,$$

etc.

As for circles or spheres (see Section 3.1), equation (3.3.27) shows that the contact points of the tangents of a conic (quadric) through M_0 belong to the straight line (plane, respectively) of equation (3.3.28) (because these contact points must satisfy $f(\bar{r}) = 0$). Accordingly, this line (plane) has an important geometric role even if M_0 is not on Γ. It is called the *polar line (plane)* of the point M_0 with respect to Γ, and the point is the *pole* of the corresponding line (plane). We advise the reader to look again at Example 3.1.6 in order to understand how to look for the pole of a given line (plane).

Another interesting notion is introduced by

3.3.12 Definition. *The straight line which passes through the contact point, and is perpendicular to the tangent line (plane) is called the normal of the conic (quadric) at the given contact point.*

Finally, if two conics (quadrics) have a common point, M then, by definition, their angle at M is the angle between either the tangents (tangent planes) or the normals of the conics (quadrics) at M. Notice that, if there are more common points, the angles may be different at different common points.

We end the discussion about tangents by

3.3.13 Remark. *For a fixed direction defined by a vector $\bar{v}$, and for a given conic (quadric) Γ, the set of tangents of Γ which are parallel to $\bar{v}$ is represented by the equation*

$$B^2(\bar{r},\bar{v}) - (\bar{v}.T\bar{v})f(\bar{r}) = 0, \tag{3.3.30}$$

where $\bar{r}$ is a generic radius vector.

Indeed, this is exactly the tangency equation (3.3.22) for the direction $\bar{v}$. Obviously, for a conic we have two tangents parallel to $\bar{v}$ (equation (3.3.30) is quadratic), and for a quadric we have a tangent (quadratic) cylinder with generators parallel to $\bar{v}$.

Another application of the intersection equation (3.3.13) is that of finding centers of symmetry of a conic (quadric). A point M_0 is a *symmetry center, briefly, a center* of the conic (quadric) Γ defined by equation (3.3.6) if, for any straight line d through M_0, M_0 is the midpoint of the segment M_1M_2, where M_1, M_2 are the intersection points of d with Γ.

3.3.14 Proposition. *M_0 is a symmetry center of Γ iff*

$$T\bar{r}_0 + \bar{a} = 0, \tag{3.3.31}$$

where the notation is that of (3.3.6).

Proof. From equation (3.3.12) of d, it follows that M_0 is a center of symmetry iff, for any vector $\bar{v} \neq \bar{0}$, the solutions λ_1, λ_2 of the intersection equation (3.3.13) satisfy the condition $\lambda_1 + \lambda_2 = 0$ (to see this, use a unit vector $\bar{v}$). The relations between the coefficients and the solutions of (3.3.13) tell us that this happens iff (3.3.31) holds. Q.E.D.

It is important to note that the notion of symmetry center has an affine character, and, with respect to affine coordinates, we find the symmetry centers using the matrix version of (3.3.31) in agreement with our conventions at the beginning of this section, namely

$$A\xi + a = 0, \tag{3.3.31'}$$

where the notation is that of (3.3.7), (3.3.8), (3.3.9).

A system of linear equations (3.3.31′) may have either a unique solution, or infinitely many solutions, or no solution. Hence, the same can be said for the center of symmetry of a conic (quadric). In a more precise way, the study of the solutions of (3.3.31′) ("by hand" or, better, by using theorems about systems of linear equations) shows that Γ has a unique symmetry center iff

$$\delta := det(A) \neq 0. \tag{3.3.32}$$

If $\delta = 0$, Γ has either infinitely many centers or no center. The general theorems about linear systems of equations (e.g., Kronecker-Capelli) show that centers exist iff the matrices A and $\tilde{A}$ have the same rank. In particular, if $\delta = 0$ but

$$\Delta := det(\tilde{A}) \neq 0, \tag{3.3.33}$$

Γ has no center. On the other hand, infinitely many centers may exist iff $\Delta = 0$. (The significance of this condition will be discovered in the next section.)

We ask the reader to study the centers of the conics and quadrics defined by the canonical equations of Section 3.2, and find again the symmetry centers indicated there. On the other hand, for example, a pair of parallel straight lines d_1, d_2 is a conic with infinitely many centers (all the points of $d||d_1, d_2$ at equal distance from d_1, d_2). The cylinder

$$\frac{x^2}{a^2} \pm \frac{y^2}{b^2} - 1 = 0;$$

is an example of a quadric with a line of centers namely, the z-axis. A pair of parallel planes, obviously is a quadric with a plane of centers.

If a conic (quadric) Γ has a center, which is not on Γ, the straight lines through the center, and with asymptotic directions are asymptotes of Γ. (See (3.3.16).)

3.3.15 Proposition. *With respect to a frame whose origin is a center of the conic (quadric) Γ, the equation of Γ has no terms of the first degree in x, y, z.*

Proof. The equation of Γ must be of the second degree, and it must be satisfied simultaneously for (x, y, z) and $(-x, -y, -z)$. Q.E.D.

Notice the following consequence of Proposition 3.3.15 and of the remark which precedes it: the equation of the asymptotes of a conic (quadric) through a symmetry center is given by the annulation of the group of terms of the second degree in the equation of the conic (quadric) with respect to a frame with origin at that center.

A similar use of the intersection equation (3.3.13) as for centers allows us to prove

3.3.16 Proposition. *Let $\bar{v} \neq \bar{0}$ define a fixed direction. Then, if nonvoid, the geometric locus of the midpoints of the segments defined by the (real or imaginary) intersection points of all the straight lines of direction $\bar{v}$ with a given conic (quadric) Γ is a straight line (respectively, a plane), called the conjugated diameter (conjugated diametral plane) of the direction $\bar{v}$.*

Proof. As in the explanation given for centers in the proof of Proposition 3.3.14, and by looking at the intersection equation (3.3.13), it follows that a point M_0 belongs to the required locus iff

$$\bar{r}_0.T\bar{v} + \bar{a}.\bar{v} = 0. \tag{3.3.34}$$

If $T\bar{v} = \bar{0}$, either the locus is void or the problem makes no sense. The last situation is for $T\bar{v} = \bar{0}$, $\bar{a}.\bar{v} = 0$; then, equation (3.3.13) reduces to $f(\bar{r}_0) = 0$, and either it has no solution or it has infinitely many solutions, and we cannot speak of the midpoint. If $T\bar{v} \neq \bar{0}$, $\bar{r}_0$ of (3.3.34) is the radius vector of a generic point of coordinates (x, y, z) of the locus, and the locus is the line (plane) of equation (3.3.34). (We may replace $\bar{r}_0$ by $\bar{r}$ in this equation.) Q.E.D.

The name *diameter* or *diametral plane* comes from the fact shown by equations (3.3.34) and (3.3.31) that, if the conic (quadric) has a center, the line (plane) (3.3.34) contains the center.

3.3.17 Definition. *Any direction $\bar{w}$ which is parallel with the conjugated diameter (diametral plane) of a given direction $\bar{v}$ is called a conjugated direction of $\bar{v}$.*

3.3.18 Proposition. *$\bar{w}$ is conjugated to $\bar{v}$ iff $\bar{w}.T\bar{v} = 0$, and, then, $\bar{v}$ is conjugated to $\bar{w}$.*

Proof. Equation (3.3.34) shows that $\bar{w}$ is parallel to the line (plane) (3.3.34) iff $\bar{w}.T\bar{v} = 0$. The second assertion (i.e., the symmetry of the conjugation relation) follows from Lemma 3.3.1. Q.E.D.

Proposition 3.3.18 also shows that the asymptotic directions are characterized by the fact that they are *self-conjugated* (conjugated to themselves).

Our next subject will be that of symmetry axes and symmetry planes of conics and quadrics. These depend on orthogonality hence, we use orthonormal frames to study them.

3.3.19 Definition. *A direction $\bar{v}$ which is orthogonal to all its conjugated directions is called a principal direction of the conic (quadric), and the conjugated diameter (diametral plane) of such a direction is called a principal diameter (plane).*

Thus, a principal diameter of a conic divides its orthogonal chords into equal parts and, consequently, it is a *symmetry axis* of the conic (see earlier Figs. 3.2.1, 3.2.2). For the same reason, a principal plane of a quadric is a *plane of symmetry* (Figs. 3.2.6, 3.2.7, 3.2.8, 3.2.9, 3.2.10). Obviously, the symmetry axes and planes are very important for the geometry of the conics (quadrics), and we will study the procedure to find them.

3.3.20 Proposition. *A vector $\bar{v} \neq \bar{0}$ defines a principal direction of a* *conic (quadric)* Γ *iff there exists a number* s *such that*

$$T\bar{v} = s\bar{v}. \tag{3.3.35}$$

Furthermore, the conjugated diameter (plane) of $\bar{v}$ *is a symmetry axis (plane) iff* s *is real and* $s \neq 0$.

Proof. By Proposition 3.3.18, $T\bar{v}$ is perpendicular to all the conjugated vectors of $\bar{v}$. Since, now, we want $\bar{v}$ to have the same property, we must have (3.3.35), and conversely. Since this direction $\bar{v}$ has a conjugated diameter (plane) iff $T\bar{v} \neq \bar{0}$, the last assertion of our proposition is also clear. Q.E.D.

We will also refer to vectors $\bar{v}$ which satisfy (3.3.35) as *principal vectors.* For the same vectors $\bar{v}$, another name is provided by *linear algebra: eigenvectors* (which is German for *self-vectors*) of the operator T. The value of s of (3.3.35) to which the eigenvector $\bar{v}$ corresponds is called the corresponding *eigenvalue* of the operator T. Clearly, an eigenvector corresponds to a uniquely defined eigenvalue but, we may have many eigenvectors associated to the same eigenvalue.

We must handle equation (3.3.35) in such a way as to determine both s and $\bar{v}$ from it. The key for this is the fact that the eigenvectors are different from $\bar{0}$.

In the case of a conic, equation (3.3.35) is equivalent to

$$(a_{11} - s)v_1 + a_{12}v_2 = 0, \tag{3.3.36}$$
$$a_{21}v_1 + (a_{22} - s)v_2 = 0.$$

In the case of a quadric, the corresponding system of equations is

$$\begin{aligned} &(a_{11} - s)v_1 + a_{12}v_2 + a_{13}v_3 = 0, \\ &a_{21}v_1 + (a_{22} - s)v_2 + a_{23}v_3 = 0, \\ &a_{31}v_1 + a_{32}v_2 + (a_{33} - s)v_3 = 0. \end{aligned} \tag{3.3.37}$$

It is well known that (3.3.36) has *nontrivial solutions* $\bar{v} \neq \bar{0}$ iff

$$\begin{vmatrix} a_{11} - s & a_{12} \\ a_{21} & a_{22} - s \end{vmatrix} = 0. \tag{3.3.38}$$

Similarly, the system (3.3.37) has nontrivial solutions $\bar{v} \neq 0$ iff

$$\begin{vmatrix} a_{11}-s & a_{12} & a_{13} \\ a_{21} & a_{22}-s & a_{23} \\ a_{31} & a_{32} & a_{33}-s \end{vmatrix} = 0. \tag{3.3.39}$$

The equations (3.3.38), (3.3.39) are called *the characteristic equation* (or, with an older term, *the secular equation*), and the solutions of this equation are the eigenvectors of the operator T.

Furthermore, for each eigenvalue, we will write down the corresponding system (3.3.36), (3.3.37). The result is called the *characteristic system* of T for that eigenvalue, and the nontrivial solutions $\bar{v}$ of the characteristic system are the principal vectors (directions) of the conic (quadric). Finally, the conjugated diameters (planes) of these vectors are the symmetry axes (planes) of the conic (quadric).

In order to show how useful the symmetry axes (planes) can be, we prove

3.3.21 Proposition. *If $x = 0$ is a symmetry axis (plane) of a conic (quadric) Γ, then, either Γ contains this axis (plane) or the equation of Γ contains no term where x enters at degree* 1. *(Similar results hold for the coordinates y, z.)*

Proof. Suppose that the equation of Γ is (3.3.1). If $x = 0$ is an axis of symmetry, (x, y) and $(-x, y)$ must satisfy (3.3.1) simultaneously; this happens iff $\forall(x, y) \in \Gamma$, we have simultaneously

$$a_{11}x^2 + a_{22}y^2 + 2a_{20}y + a_{00} = 0, \quad a_{12}xy + a_{10}x = 0. \tag{3.3.40}$$

Therefore, the equation of Γ is either the first or the second equation (3.3.40), and the coefficients of the other equation are zero. Equation (3.3.2) may be discussed similarly. Q.E.D.

Another important result is

3.3.22 Proposition. *Every conic (quadric) has at least one symmetry axis (plane).*

Proof. For a conic, the characteristic equation (3.3.38) is

$$s^2 - (a_{11} + a_{22})s + \delta = 0, \tag{3.3.41}$$

where δ was defined in (3.3.32). This equation has real solutions because

$$(a_{11}+a_{22})^2 - 4(a_{11}a_{22}-a_{12}^2) = (a_{11}-a_{22})^2 + 4a_{12}^2 \geq 0.$$

Furthermore, both solutions of (3.3.41) are 0 iff the equation reduces to $s^2 = 0$ i.e.,

$$a_{11}+a_{22}=0, \; a_{11}a_{22}-a_{12}^2=0,$$

or, equivalently,

$$a_{22}=-a_{11}, \; a_{11}^2+a_{12}^2=0.$$

Since the coefficients a_{ij} are real numbers, the previous condition holds iff $a_{11}=a_{22}=a_{12}=0$, which is not the case of a conic.

In the case of a quadric, the characteristic equation (3.3.39) is of the third degree, and, since its coefficients are real (hence the complex solutions, if any, come in conjugated pairs), we must have at least one real eigenvalue. If this eigenvalue is not zero, we are done. If it is zero, and if $\bar{v}$ is a corresponding eigenvector, we will change the frame such that the new z-axis has direction $\bar{v}$. This will not influence our results since the definition of principal directions and of symmetry axes and planes was geometric, independent of the choice of a frame. For the new frame, $(0,0,1)$ is a solution of the system (3.3.37) with $s=0$, whence $a_{13}=a_{23}=a_{33}=0$. Now, if we look again at the characteristic equation (3.3.39), we see that for the other eigenvalues, we remain with exactly the same equation (3.3.38) as in the case of a conic. Hence, a nonzero real solution will exist, as for conics.

Now, our proposition follows from Proposition 3.3.21. Q.E.D.

We ask the reader to use the present methods in order to prove that the symmetry axes and planes of the nondegenerate conics and quadrics are exactly those indicated in Section 3.2.

The problems discussed in the present section are only the most important geometric problems concerning conics and quadrics defined by general equations. Many more problems could be studied. For instance, two generic conics intersect at four points, two generic quadrics intersect following a more complicated curve called a *quartic.* There exists a general theory of focal points, etc. We will only mention that we can define the notion of a *pencil of conics (quadrics)*, and get the corresponding equation, in exactly the same way as for circles and spheres (see Section 3.1).

Exercises and Problems

3.3.1. Find a tangent of the hyperbola $x^2/9 - y^2/16 - 1 = 0$ which is at equal distance from the center and from one of the focuses of the hyperbola.

3.3.2. Find the common tangents of the ellipse $x^2/45 + y^2/20 - 1 = 0$, and the parabola $y^2 - (20/3)x = 0$.

3.3.3. Find the equation of a conic which passes through the origin, is tangent to $4x + 3y + 2 = 0$ at $(1, -2)$, and is tangent to $x - y - 1 = 0$ at $(0, -1)$.

3.3.4. Find the equations of the sides of a square circumscribed to the ellipse $x^2/6 + y^2/3 - 1 = 0$.

3.3.5. Prove that any tangent of a hyperbola meets the asymptotes at two points which are symmetric with respect to the contact point.

3.3.6. Let Γ be a conic, M_0 a fixed point of Γ, and A, B points which vary on Γ such that $AM_0 \perp BM_0$. Prove that the line AB passes through a fixed point situated on the normal of Γ at M_0.

3.3.7. Prove that if an ellipse and a hyperbola have the same focuses the angle between these two curves at their intersection points is a right angle.

3.3.8. Find the locus of the points M of the plane of an ellipse, hyperbola or parabola Γ such that the tangents of Γ through M are orthogonal.

3.3.9. Let Γ be a parabola of focus F and directrix δ. Let $M \in \Gamma$, $d =$ the tangent of Γ at M, and $N = d \cap \delta$. Find the locus of the intersection of MF with the parallel to the axis of Γ through N.

3.3.10. Prove that the polar lines of a fixed point P with respect to all the conics of a pencil of conics form a pencil of lines.

3.3.11. Show that the family of conics which contain $A(1, 0)$, $B(0, 1)$ and are tangent at O to the bisector of the second coordinate quadrant is a pencil of conics. Find the axes of these conics.

3.3.12. Let xOy be a plane, orthogonal frame, and $A \in Ox$. Consider all the conics through A, tangent to Oy at O, and with orthogonal asymptotic directions. Find the locus of the centers of these conics.

3.3.13. Prove that if a rectangle has its vertices on a conic Γ, the sides of the rectangle must be parallel to the axes of Γ.

3.3.14. Write down the general equation of all the conics of a given center (x_0, y_0).

3.3.15. Prove that the polar line of any point of an asymptote of a hyperbola is parallel to the asymptote.

3.3.16. Change the coordinates (x, y) in such a way that the new axes are the asymptotes of the conic Γ defined by

$$2x^2 - 7y^2 - 12xy + 8x + 6y = 0.$$

Write down the equation of Γ with respect to this new frame.

3.3.17. Find the general equation of the conic which has the lines

$$ax + by + c = 0, \quad a'x + b'y + c' = 0$$

as its asymptotes.

3.3.18. Let $\mathcal{H}$ be a hyperbola, and let $ABCD$ be a parallelogram with $A, C \in \mathcal{H}$, and whose sides are parallel to the asymptotes of $\mathcal{H}$. Prove that the line BD passes through the center of the hyperbola.

3.3.19. Find the midpoint of the chord of intersection of

$$2x^2 + 3y^2 + 4xy - 3x - 3y = 0$$

with the line $x + 3y - 12 = 0$.

3.3.20. Find the diameter of the conic

$$2x^2 + 9y^2 - 6xy - 12x + 14y - 7 = 0$$

which makes an angle of $\pi/4$ with its conjugate direction.

3.3.21. Let Γ be a conic with center C, and let M_0 be a point in the plane of the conic. Let A, B be the contact points of the tangents of

Γ through M_0. Prove that the line M_0C cuts the chord AB into equal parts.

3.3.22. Find the axes of the conic

$$5x^2 - 2y^2 + 24xy + 4x - 1 = 0.$$

3.3.23. Write down the equation of the normal of the ellipsoid

$$\frac{x^2}{8} + \frac{y^2}{4} + \frac{z^2}{1} = 1$$

at $(2, 1, -1/2)$.

3.3.24. Find the tangent planes of

$$\frac{x^2}{21} + \frac{y^2}{6} + \frac{z^2}{4} = 1$$

which are parallel to the plane $2x + 2y - 3z = 0$.

3.3.25. Find the tangent planes of

$$\frac{x^2}{36} + \frac{y^2}{9} - \frac{z^2}{4} = 1$$

which pass through the line

$$\frac{x}{6} = \frac{y}{-3} = \frac{z+2}{4}.$$

3.3.26. Write down the equation of a quadric which cuts the plane xOy by a parabola, and the planes xOz, yOz by circles of radius r, tangent to the positive halves of the coordinate axes.

3.3.27. Let Γ be a quadratic cone, and g_1, g_2 two of its generator lines. Prove that the tangent plane of Γ along any generator is a fixed plane. If π_{g_1}, π_{g_2} are the tangent planes of Γ along g_1, g_2, show that the diametral plane conjugated with the direction $\pi_{g_1} \cap \pi_{g_2}$ is the plane (g_1, g_2).

3.3.28. Write down the equation of a quadric which has the center $C(0,0,1)$, contains the point $M(2,0,-1)$, and cuts the plane $z = 0$ by the curve $x^2 - 4xy - 1 = 0$, $z = 0$.

3.3.29. Write down the equation of the cone of asymptotic directions at $(1,-1,3)$ for the quadric

$$4x^2+6y^2+4z^2+4xz-8y-4z+3=0.$$

3.3.30. Find the diameter of the quadric

$$x^2+9y^2+2z^2-4xy-6xz+2yz+8x-16y+1=0$$

which passes through the origin, and its conjugated diametral plane.

3.3.31. Find the rectilinear generators of the paraboloid $x^2/16-y^2/4=z$ which are parallel to the plane $3x+2y-4z=0$.

3.3.32. Find the symmetry axes of the quadric

$$x^2+y^2+5z^2-6xy-2xz+2yz-6x+6y-6z+9=0.$$

3.3.33. Find the symmetry planes of the quadric

$$x^2-2y^2+z^2+4xy-8xz-4yz-14x-4y+14z+18=0,$$

and prove that this is a quadric of revolution.

3.3.34. Find the equation of the quadric which contains Oz, its tangent plane at O is $x=0$, the directions of Ox, Oy are asymptotic directions and the polar plane of $(1,1,1)$ is $x+y+z+2=0$.

3.3.35. Let Γ be a quadric with center C, and π a pencil of parallel planes. Prove that the locus of the centers of the conics $\pi\cap\Gamma$ is the conjugated diameter of the diametral plane of Γ contained in π.

3.3.36. Let Γ be a quadric. Prove that, if Γ has two parallel, tangent planes π_1, π_2, the chord d which joins their contact points is a diameter, and, conversely, the tangent planes at the ends of a diameter are parallel.

3.3.37. Find the planes which are tangent to the paraboloid $x^2/4+y^2/2=z$, and such that any parallel plane cuts the paraboloid by circles. (Hint: First, find the planes which cut the paraboloid by circles using interior coordinates of such planes.)

3.4 Classifications of Conics and Quadrics

After having studied the general geometric properties of conics and quadrics, we can address the problem of their classification. This will tell us exactly what geometric objects are represented by a quadratic equation.

In principle, the classification is done by establishing the different, simplest forms which a quadratic equation can take after a convenient choice of the frame. These simplest forms are called the *canonical equations*, and include the canonical equations encountered in section 3.2. From the very definition of the classification problem, it follows that we will have to discuss the *orthogonal classification* and the *affine classification* of conics and quadrics, according to the type of frames which we are allowed to use.

We begin with the orthogonal classification.

Let Γ be a conic defined by the general equation (3.3.1). By Proposition 3.3.22, we can find a symmetry axis of Γ, and we may use it as the x-axis of a new orthogonal frame. Then, by Proposition 3.3.21, the equation of Γ becomes

$$a_{11}x^2 + a_{22}y^2 + 2a_{10}x + a_{00} = 0. \tag{3.4.1}$$

Now, the following cases appear:

1) The x-axis has 2 imaginary, conjugated intersection points with Γ. Then, if the origin is taken at the corresponding midpoint, equation (3.4.1) takes the form

$$x^2 + a_{22}y^2 + m^2 = 0, \qquad (m \in \mathbf{R}) \tag{3.4.2}$$

and a comparison with the canonical equations of section 3.2 shows that we may have the following objects, exactly:
1i) an *imaginary ellipse* $x^2/a^2 + y^2/b^2 + 1 = 0$ ($a^2 := m^2$, $b^2 := m^2/a_{22}$, $a_{22} > 0$) (it has no real points);
1ii) a hyperbola $x^2/a^2 - y^2/b^2 + 1 = 0$ ($a^2 := m^2$, $b^2 := -m^2/a_{22}$, $a_{22} < 0$);
1iii) a pair of parallel, imaginary straight lines $x^2 + m^2 = 0$ ($a_{22} = 0$).

2) The x-axis has 2 real intersection points with Γ. Then, if the origin is taken at the corresponding midpoint, equation (3.4.1) takes

the form

$$(3.4.3) \qquad x^2 + a_{22}y^2 - m^2 = 0, \qquad (m \in \mathbf{R})$$

and this equation includes:
2i) an *ellipse* $x^2/a^2 + y^2/b^2 - 1 = 0$ ($a^2 := m^2$, $b^2 := m^2/a_{22}, a_{22} > 0$);
1ii) a hyperbola $x^2/a^2 - y^2/b^2 - 1 = 0$ ($a^2 := m^2$, $b^2 := -m^2/a_{22}$, $a_{22} < 0$);
1iii) a pair of parallel, real straight lines $x^2 - m^2 = 0$.

3) The x-axis has 2 coincident (necessarily real) intersection points with Γ. Then, if the origin is taken at the intersection point, equation (3.4.1) takes the form

$$(3.4.4) \qquad x^2 + a_{22}y^2 = 0,$$

with the following three subcases:
3i) a pair of real, intersecting straight lines, if $a_{22} < 0$;
3ii) a pair of imaginary, intersecting straight lines, if $a_{22} > 0$;
3iii) a pair of coincident lines $x^2 = 0$, if $a_{22} = 0$.

4) The x-axis has an asymptotic direction. Then, if the origin is taken at the intersection point of the x-axis with Γ, equation (3.4.1) takes the form

$$(3.4.5) \qquad a_{22}y^2 + 2a_{10}x = 0, \qquad (a_{10} \neq 0)$$

and Γ is a parabola $y^2 - 2px = 0$, ($p = -a_{10}/a_{22}$). Notice that we must have $a_{22} \neq 0$ since, otherwise, we do not have a quadratic equation.

5) The x-axis is an asymptote. Then, (3.4.1) must be of the form $y^2 + t = 0$ ($t \neq 0$), and, again, we have a pair of parallel lines (real or imaginary).

6) The x-axis belongs to the conic. Then, (3.4.1) must reduce to $y^2 = 0$, and Γ is a pair of coincident straight lines.

The result of this discussion is the following *classification theorem*:

3.4.1 Theorem. *The orthogonal classification of conics, and their canonical equations are as follows:*
1) imaginary ellipses: $x^2/a^2 + y^2/b^2 + 1 = 0$;
2) real ellipses: $x^2/a^2 + y^2/b^2 - 1 = 0$;
3) hyperbolas: $x^2/a^2 - y^2/b^2 - 1 = 0$;

4) parabolas: $y^2 - 2px = 0$;
(All these are called nondegenerate conics.)
5) pairs of real, intersecting, straight lines: $x^2 - a^2y^2 = 0$;
6) pairs of imaginary, intersecting, straight lines: $x^2 + a^2y^2 = 0$;
7) pairs of real, parallel, straight lines: $x^2 - a^2 = 0$;
8) pairs of imaginary, parallel, straight lines: $x^2 + a^2 = 0$;
9) pairs of coincident (real), straight lines: $x^2 = 0$.
(In the last five cases, the conic is a degenerate conic.)
The orthogonal type of a conic is uniquely defined. The numerical coefficients which enter in the canonical equations above are real numbers, and have a geometric meaning; i.e., two conics of the same orthogonal type are distinguished by the values of these coefficients. The same is true for the corresponding canonical orthonormal frames, i.e., these frames have a geometric meaning.

Only the last assertions deserve a few more words of explanation. Namely, as seen in the classification process, the canonical frame, the coefficients, and the orthogonal type of the conic were determined by various *geometric properties* (symmetry axes and their intersection with the conic). Hence, they depend on the conic and not on the orthogonal frame used initially. The coefficients of the canonical equations are called *orthogonal invariants* of the conic. Notice that the canonical frame of a conic may not be unique. Indeed, it is possible to change the role of the coordinate axes e.g., for an ellipse changing (x, y) by $(\tilde{x} = -y, \tilde{y} = x)$ doesn't alter the canonical character of the equation. And, in the case of a circle, we have infinitely many canonical frames: any orthogonal frame with the origin at the center is canonical.

The classification procedure described above also has a practical value. Given the general equation of a conic, we know how to actually look for symmetry axes (section 3.3). Therefore, we can find the concrete canonical frame and equation of the conic. Practically, if the conic has a center, it is convenient to translate the origin to the center first, and use the symmetry axes, afterwards.

Now, let us go over to the quadrics. For a quadric Γ, defined by the general equation (3.3.2), we can always find a symmetry plane (Proposition 3.3.22). If we choose this plane to be $z = 0$, Proposition 3.3.21 tells us that the new equation of Γ contains no term where z has exponent

1. The intersection of the plane $z = 0$ with Γ is a conic; hence, working in the (x, y)-plane, we can find the canonical x, y-axes of the intersection conic, while the z-axis will be the normal of the (x, y)-plane at the canonical origin. Therefore, with respect to this frame, the equation of the quadric Γ reduces to the canonical equation of the intersection conic, plus a term in z^2. The discussion of this new equation, and the canonical equations of quadrics given in section 3.2, provide us with the classification of the quadrics.

The possible cases are

1) $x^2/a^2 + y^2/b^2 + a_{33}z^2 + 1 = 0$, with the subcases:
1i) an *imaginary ellipsoid* (because it contains no real points), if $a_{33} > 0$;
1ii) a *two-sheeted hyperboloid*, if $a_{33} < 0$;
1iii) an *imaginary elliptic cylinder* because the equation has only two variables; see Proposition 2.3.9), if $a_{33} = 0$.

2) $x^2/a^2 + y^2/b^2 + a_{33}z^2 - 1 = 0$, with the subcases:
2i) a *real ellipsoid*, if $a_{33} > 0$;
2ii) a *one-sheeted hyperboloid*, if $a_{33} < 0$;
2iii) a *real elliptic cylinder*, if $a_{33} = 0$.

3) $x^2/a^2 - y^2/b^2 + a_{33}z^2 - 1 = 0$, with the subcases:
3i) a *one-sheeted hyperboloid*, if $a_{33} > 0$;
3ii) a *two-sheeted hyperboloid*, if $a_{33} < 0$;
3iii) a *hyperbolic cylinder*, if $a_{33} = 0$.

4) $y^2 + a_{33}z^2 - 2px = 0$, with the subcases:
4i) an *elliptic paraboloid*, if $a_{33} > 0$;
4ii) a *hyperbolic paraboloid*, if $a_{33} < 0$;
4iii) a *parabolic cylinder*, if $a_{33} = 0$.

5) $x^2 - a^2y^2 + a_{33}z^2 = 0$, with the subcases:
5i) a *real quadratic cone*, if either $a_{33} > 0$ or $a_{33} < 0$ (see again Proposition 2.3.9);
5ii) a pair of real intersecting planes, if $a_{33} = 0$.

6) $x^2 + a^2y^2 + a_{33}z^2 = 0$, with the subcases:
6i) an *imaginary cone*, if $a_{33} > 0$;
6ii) a *real cone*, if $a_{33} < 0$;
6iii) a pair of imaginary intersecting planes, if $a_{33} = 0$.

The reader will see by himself, now, that the not yet encountered types of quadrics which are found from the cases 7,8,9 of Theorem 3.4.1 are pairs of parallel planes (real and imaginary), and pairs of coincident planes.

The above discussion of quadrics allows us to give the following brief formulation of the *orthogonal classification theorem of quadrics*

3.4.2 Theorem. *The orthogonal classification of quadrics consists of the following classes: 1) the nondegenerate quadrics : ellipsoids (real and imaginary), hyperboloids (one and two-sheeted), paraboloids (elliptic and hyperbolic), with the canonical equations given in section 3.2; 2) degenerate quadrics: various types of quadratic cones, quadratic cylinders and pairs of planes.*

The total number of the classes of Theorem 3.4.2 is 17.

Finally, let us emphasize that all that was said about the invariance of the orthogonal type of a conic, its invariant canonical frame (or frames) and invariant coefficients, and about the concrete finding of the type, and the canonical equation holds for quadrics as well.

3.4.3 Example. Find the canonical, orthogonal equation, and the type of the quadric

$$x^2 + y^2 - 3z^2 - 2xy - 6xz - 6yz + 2x + 2y + 4z = 0.$$

Solution. The characteristic equation is

$$\begin{vmatrix} 1-s & -1 & -3 \\ -1 & 1-s & -3 \\ -3 & -3 & -3-s \end{vmatrix} = (2-s)(s^2+3s-18) = 0,$$

and the eigenvalues are $s_1 = 2$, $s_2 = 3$, $s_3 = -6$.

For the eigenvalue s_1, the characteristic system is

$$\begin{aligned} -v_1 - v_2 - 3v_3 &= 0, \\ -v_1 - v_2 - 3v_3 &= 0, \\ -3v_1 - 3v_2 - 5v_3 &= 0, \end{aligned}$$

and the corresponding principal direction is $\bar{v}_1(1,-1,0)$. In the same way, we can find the principal directions associated with the other eigenvalues: $\bar{v}_2(1,1,-1)$ and $\bar{v}_3(1,1,2)$.

The conjugated diametral plane of the direction $\bar{v}_1$ is $\bar{r}.T\bar{v}_1 + \bar{a}.\bar{v}_1 = 0$, which is

$$x - y = 0.$$

As we know, this plane is a symmetry plane of the quadric. Similarly, we have two more symmetry planes, conjugated with $\bar{v}_2$, $\bar{v}_3$:

$$x + y - z = 0, \; 3x + 3y + 6z - 2 = 0.$$

Now, we want to go over to a new orthogonal frame where $\bar{v}_1$ is the direction of the new z-axis. But, as we said in the general explanations, it is a good idea to look for the center of the quadric first. If there is no center, we manage without it, and go straight to the next phase. The center is $C(x_0, y_0, z_0)$, where the coordinates are the solutions of the system

$$\begin{aligned} x_0 - y_0 - 3z_0 + 1 &= 0, \\ -x_0 + y_0 - 3z_0 + 1 &= 0, \\ -3x_0 - 3y_0 - 3z_0 + 2 &= 0, \end{aligned}$$

which gives $x_0 = 1/6, \; y_0 = 1/6, \; z_0 = 1/3$. (We could have checked in advance that $\delta \neq 0$, and, thereby, show that the quadric has a unique center.)

We use the center by making the translation

$$\tilde{x} = x - \frac{1}{6}, \; \tilde{y} = y - \frac{1}{6}, \; \tilde{z} = z - \frac{1}{3},$$

which leads to a frame with the origin at the center. The equation of the quadric with respect to these new coordinates is

$$\tilde{x}^2 + \tilde{y}^2 - 3\tilde{z}^2 - 2\tilde{x}\tilde{y} - 6\tilde{x}\tilde{z} - 6\tilde{y}\tilde{z} + 1 = 0.$$

Of course, the translation leaves the coordinates of the principal vectors unchanged. We need an orthonormal basis $(\bar{i}', \bar{j}', \bar{k}')$ such that $\bar{k}' = \bar{v}_1/|\bar{v}_1|$. For this purpose, we can first find a solution $\bar{w}(w_1, w_2, w_3)$ of length 1 of the equation $\bar{v}_1.\bar{w} = 0$, $\bar{w}$ i.e., $w_1 - w_2 = 0$, and take this solution as $\bar{i}'$. Then, we define $\bar{j}' = \bar{k}' \times \bar{i}'$. It happens, (and not accidentally!) that the other two principal vectors are orthogonal to $\bar{v}_1$, and it will be convenient to take

$$\begin{aligned} \bar{i}' &= \frac{1}{\sqrt{3}}(\bar{i} + \bar{j} - \bar{k}), \\ \bar{j}' &= \frac{1}{\sqrt{6}}(\bar{i} + \bar{j} + 2\bar{k}), \end{aligned}$$

$$\bar{k}' = \frac{1}{\sqrt{2}}(\bar{i} - \bar{j}).$$

The reason for this choice is that all the new coordinate planes will be planes of symmetry, and we do not have to handle separately the problem of writing the canonical equation of the conic of intersection between the quadric and the symmetry plane orthogonal to $\bar{k}'$.

Using the formula (1.3.10), we get the coordinate transformation formulas

$$\tilde{x} = \frac{1}{\sqrt{3}}x' + \frac{1}{\sqrt{6}}y' + \frac{1}{\sqrt{2}}z',$$

$$\tilde{y} = \frac{1}{\sqrt{3}}x' + \frac{1}{\sqrt{6}}y' - \frac{1}{\sqrt{2}}z',$$

$$\tilde{z} = -\frac{1}{\sqrt{3}}x' + \frac{2}{\sqrt{6}}y',$$

where (x', y', z') are the new coordinates. With respect to these new coordinates, the equation of the quadric becomes

$$3x'^2 - 6y'^2 + 2z'^2 + 1 = 0.$$

After a new change of coordinates: $x'' = y'$, $y'' = z'$, $z'' = x'$, the equation of the quadric becomes

$$\frac{x''^2}{\frac{5}{18}} - \frac{y''^2}{\frac{5}{6}} - \frac{z''^2}{\frac{5}{9}} - 1 = 0,$$

and we see that the quadric which we studied is a two-sheeted hyperboloid.

Such computations may be tedious but, they are nothing more than elementary algebra. Later on, we will see that there exist quicker ways for finding the nature (type) of a conic, but the previous technique is still needed (with shortcuts, perhaps) if we want a canonical frame.

Now, we will address the problem of the affine classification. Technically, this classification can be obtained in a simple way, by using a method (due to Gauss), which consists of building squares of linear expressions in the equation of the conic (quadric). It is simpler to explain this method on examples, which, however, will clearly show that the method is general.

3.4.4 Examples. *i).* We begin by the quadric studied in Example 3.4.3. In it, we may create a square from all the terms which contain x if we add and subtract the missing part of the square. The result is

$$[(x-y-3z+1)^2-(y^2+9z^2+1+6yz-2y-6z)]+y^2-3z^2-6yz+2y+4z=0,$$

i.e.,

$$(x-y-3z+1)^2-12z^2-12yz+10z-1=0.$$

Using, similarly, the z^2-term, we put our equation under the following form

$$(x-y-3z+1)^2-12(z+\frac{1}{2}y-\frac{5}{12})^2+3y^2-y+\frac{13}{12}=0.$$

And, finally, we may put the equation under the form

$$(x-y-3z+1)^2-12(z+\frac{1}{2}y-\frac{5}{12})^2+3(y-\frac{1}{6})^2+1=0.$$

This equation suggests the following affine change of coordinates:

$$x'=x-y-3z+1,$$
$$y'=\sqrt{3}(y-\frac{1}{6}),$$
$$z'=\sqrt{12}(z+\frac{1}{2}y-\frac{5}{12}).$$

The reader is asked to check that this is a correct affine change of coordinates (see section 1.3). After this change, the equation of the quadric becomes

$$x'^2+y'^2-z'^2+1=0,$$

and a quadric which has such an affine equation will be called an *affine two-sheeted hyperboloid*.

ii). There is a case which is not covered by the above example namely, the case where the original equation (or one of the subsequent equations) contains no square term e.g.,

$$xy+yz+zx-5=0.$$

In this case, we make a preliminary affine transformation of the following type:

$$x' = x + y,\ y' = x - y,\ z' = z.$$

After that, the equation becomes

$$x'^2 - y'^2 + \ldots = 0,$$

and we can start building squares as in the previous example.

Of course, after such transformations the equation may also be of a form which includes the squares of a part of the new coordinates, while the other coordinates enter with exponent 1.

The existence of this *Gauss' method* proves the following *affine classification theorem for conics and quadrics*

3.4.5 Theorem. *Every conic Γ has an affine canonical equation, and an affine nature (type) of one of the following classes:*
1) $x^2 + y^2 + 1 = 0$ *(affine imaginary ellipse);*
2) $x^2 + y^2 - 1 = 0$ *(affine real ellipse);*
3) $x^2 - y^2 - 1 = 0$ *(affine hyperbola);*
4) $y^2 - 2x = 0$ *(affine parabola);*
5) $x^2 - y^2 = 0$ *(a pair of real concurrent straight lines);*
6) $x^2 + y^2 = 0$ *(a pair of imaginary concurrent straight lines);*
7) $x^2 - 1 = 0$ *(a pair of real parallel straight lines);*
8) $x^2 + 1 = 0$ *(a pair of imaginary parallel straight lines);*
9) $x^2 = 0$ *(a pair of coincident straight lines).*
The conics 1) to 4) are called nondegenerate; the conics 5) to 9) are degenerate.

3.4.6 Theorem. *Every quadric Γ has an affine canonical equation, and an affine nature (type) of one of the following classes:*
1) $x^2 + y^2 + z^2 + 1 = 0$ *(affine imaginary ellipsoid);*
2) $x^2 + y^2 + z^2 - 1 = 0$ *(affine real ellipsoid);*
3) $x^2 - y^2 - z^2 - 1 = 0$ *(affine two-sheeted hyperboloid);*
4) $x^2 + y^2 - z^2 - 1 = 0$ *(affine one-sheeted hyperboloid);*
5) $x^2 + y^2 - 2z = 0$ *(affine elliptic paraboloid);*
6) $x^2 - y^2 - 2z = 0$ *(affine hyperbolic paraboloid);*
7) $x^2 + y^2 + z^2 = 0$ *(imaginary cone);*
8) $x^2 + y^2 - z^2 = 0$ *(real cone);*

9) $x^2 + y^2 + 1 = 0$ (imaginary cylinder);
10) $x^2 + y^2 - 1 = 0$ (elliptic cylinder);
11) $x^2 - y^2 - 1 = 0$ (hyperbolic cylinder);
12) $y^2 - 2x = 0$ (parabolic cylinder);
13) $x^2 + y^2 = 0$ (a pair of imaginary concurrent planes;
14) $x^2 - y^2 = 0$ (a pair of real concurrent planes;
15) $x^2 - 1 = 0$ (a pair of real parallel planes);
16) $x^2 + 1 = 0$ (a pair of imaginary parallel planes);
17) $x^2 = 0$ (a pair of coincident planes).
The quadrics 1) to 6) are called nondegenerate; the quadrics 7) to 17) are degenerate.

We emphasize again that these two theorems need no other proof than to check that, logically, these are exactly all the possible outcomes of the Gauss method of completion of squares. On the other hand, the qualification *affine* in the names of the classes comes to tell us that we use affine frames. We will see soon that the affine and the orthogonal type of a given conic (quadric) are the same but, the affine canonical equations contain no invariant coefficients. For instance, for affine geometry, any two ellipses, or an ellipse and a circle, are the same, while in orthogonal geometry the ellipses differ by the length of their half-axes.

3.4.7 Proposition. *The affine type of a conic (quadric) is unique, and it is the same as the orthogonal type.*

Proof. The uniqueness problem appears because the way of using the Gauss method is not unique. First, for conics: from the affine canonical equations it follows easily that a nondegenerate conic has no rectilinear generators; hence the conic cannot be simultaneously degenerate. In the degenerate case, the types are obviously, geometrically different. But, the same is true in the nondegenerate case as well. For instance, the ellipses, hyperbolas and parabolas have a different kind of asymptotes, and, the difference between the real and imaginary ellipses is that the former have real points.

Similar arguments will prove uniqueness in the case of the quadrics also. A quadric cannot be simultaneously degenerate and nondegenerate since the nondegenerate quadrics contain no plane, and, even if they have real rectilinear generators, the latter do not form a parallel family, as for cylinders, nor a concurrent family, as for cones. (See Proposition 3.3.8.) Furthermore, in the degenerate case, the uniqueness follows from the case of conics, for cylinders and cones, and from the kind of the pair of planes. In the nondegenerate case, the ellipsoids, hyperboloids, and paraboloids are distinguished by different types of asymptotic cones; then, the real ellipsoid is distinguished from the imaginary one by the fact that it has real points, and the one-sheeted hyperboloid is distinguished from the two-sheeted one by the fact that it has real rectilinear generators.

Now, the orthogonal canonical equation of an ellipse (3.2.3) becomes the affine canonical equation by the affine coordinate change

$$x' = \frac{x}{a},\ y' = \frac{y}{b}.$$

Hence, an orthogonal ellipse is an affine ellipse. In exactly the same way, the reader will easily see that, if a conic (quadric) has a certain orthogonal type, it has the same affine type. Conversely, an affine ellipse, say, cannot be an orthogonal hyperbola since, then, it would also be an affine hyperbola, in contradiction with the uniqueness of the affine type. A similar argument settles any other comparison of types. Q.E.D.

Because of the last result, hereafter, we will speak of the nature (type) of a conic (quadric) without adding the qualification affine or orthogonal.

We end this section by explaining how to determine the nature of a conic (quadric) from its original equation, without changing the frame first.

3.4.8 Proposition. *For conics and quadrics, the annulation or nonannulation of δ and Δ are invariant by affine coordinate transformations. Furthermore, for a conic, the sign of δ is invariant, and, for a quadric, the sign of Δ is invariant by affine coordinate transformations.*

Proof. Remember that δ and Δ were defined by (3.3.32) and (3.3.33), respectively.

Now, using matrices, a transformation of affine coordinates has equations of the form (1.3.11), and, for the coordinate columns ξ, $\mathcal{X}$ of (3.3.9), such a transformation looks as follows

$$\xi = B\xi' + C,\ \mathcal{X} = M\mathcal{X}', \tag{3.4.6}$$

where the concrete coefficient matrices B, C, are not important, and

$$M = \begin{pmatrix} B & C \\ 0 & 1 \end{pmatrix}. \tag{3.4.7}$$

The only condition is that the determinant $det(B) \neq 0$, because the coordinate transformation must be invertible. From (3.4.7), we also notice that $det(M) = det(B)$.

If (3.4.6) is inserted in the equations (3.3.10) and (3.3.11), the new matrices of the conic (quadric) are found:

$$A' = B^t A B, \quad \tilde{A}' = M^t \tilde{A} M. \tag{3.4.8}$$

Since any determinant is equal to its transposed, (3.4.8) yields

$$\delta' = det^2(B)\delta, \quad \Delta' = det^2(B)\Delta. \tag{3.4.9}$$

This justifies the invariance of the (non)annulation. It also justifies the invariance of the sign if we do not take into account that the multiplication of the equation of a conic (quadric) always is a permitted operation: it doesn't change the conic (quadric). Such a multiplication leaves the sign unchanged for determinants of even order only. This is exactly the case of δ for conics, and of Δ for quadrics. Q.E.D.

Proposition 3.4.8 allows us to establish facts about the conic (quadric) simply by calculating δ, Δ on the affine canonical equation, and, then, using the results for the original equation. For instance, we get

3.4.9 Proposition. *A conic (quadric) is nondegenerate iff* $\Delta \neq 0$.

Proof. Compute Δ for the affine canonical equations given in Theorems 3.4.5 and 3.4.6. Q.E.D.

3.4.10 Proposition. *A nondegenerate conic is an ellipse iff* $\delta > 0$, *it is a hyperbola iff* $\delta < 0$, *and it is a parabola iff* $\delta = 0$.

Proof. Same as above. Q.E.D.

Because of Proposition 3.4.10, even if the conic is degenerate one uses to say that it is of elliptic type, hyperbolic type, parabolic type according to δ as above, etc.

According to these results, and recalling that we differentiated among the affine types of conics (quadrics) by means of the behaviour of the asymptotic cones and of the existence of real points, we can build the following classification tables, where all the results can be checked on the affine canonical equations, and then used on the original equation because of invariance by affine coordinate transformations:

Conic	**Affine characteristic**
Imaginary ellipse	$\Delta \neq 0,\ \delta > 0, \cap =$ imaginary
Real ellipse	$\Delta \neq 0,\ \delta > 0, \cap =$ real
Hyperbola	$\Delta \neq 0,\ \delta < 0$
Parabola	$\Delta \neq 0,\ \delta = 0$
2 Imaginary Crossing Lines	$\Delta = 0,\ \delta > 0$
2 Real Crossing Lines	$\Delta = 0,\ \delta < 0$
2 Real Parallel Lines	$\Delta = 0,\ \delta = 0, \cap =$ 2 real points
2 Imaginary Parallel Lines	$\Delta = 0,\ \delta = 0, \cap =$ imaginary
2 Coincident Lines	$\Delta = 0,\ \delta = 0, \cap =$ 1 real point

Quadric	**Affine characteristic**
Imaginary ellipsoid	$\Delta > 0, \delta \neq 0$, Imag. asympt. cone
Real ellipsoid	$\Delta < 0, \delta \neq 0$, Imag. asympt. cone
1-Sheeted Hyperboloid	$\Delta > 0, \delta \neq 0$, Real. asympt. cone
2-Sheeted Hyperboloid	$\Delta < 0, \delta \neq 0$, Real. asympt. cone
Elliptic Paraboloid	$\Delta < 0, \delta = 0$
Hyperbolic Paraboloid	$\Delta > 0, \delta = 0$

We will not write down a table for the degenerate quadrics. The reader may try to do it by himself following the same principles. In the table given above for conics, we introduced an *intersection test* denoted by $\cap$. What we meant by this test is: take a straight line $y = k$ (or $x = k$), where $k \in (-\infty, +\infty)$, and see whether there are values of k for which you obtain the number of intersection points indicated in the table.

In the table of quadrics, one has to decide about the reality of the cone of asymptotic directions at the origin. This may be done by an intersection test, again. Namely, we will study the intersection of this cone with either planes through the vertex of the cone or planes of a pencil e.g., $z = k$.

What is important about these tables is that the characteristics involved can be computed for the original equation of the conic (quadric), without any transformation of coordinates being done first.

For instance, if we look again at the quadric of Example 3.4.3 and compute, we get $\delta = -36 \neq 0$, $\Delta = -36 < 0$. It follows that the quadric must be either a real ellipsoid or a 2-sheeted hyperboloid in accordance with the type of the cone of asymptotic directions at the origin. This cone is

$$x^2 + y^2 - 3z^2 - 2xy - 6xz - 6yz = 0.$$

If we cut it by the plane $z = 0$, we get the (double) intersection line $x - y = 0$, $z = 0$, which is real. Therefore, the cone is real and, for the third time, we see that the quadric is a 2-sheeted hyperboloid.

It is worthwhile to give one more example here

3.4.11 Example. Find the nature of the conic

$$x^2 + 8y^2 - 4xy - 8x + 6y - 5 = 0.$$

Solution. A computation yields $\delta = 4$, $\Delta = -109$. Hence, the conic is either a real or an imaginary ellipse. Now, we cut it by the line $y = k \in \mathbf{R}$. The intersection equation is

$$x^2 - 4x(k+2) + (8k^2 - 6k - 5) = 0,$$

and the nature of the solutions is given by its *discriminant* $D = -4k^2 + 22k + 21$. Clearly, there exist values of k where $D > 0$ (e.g., $k = 0$). Hence, the conic has real points, and it is a real ellipse.

Exercises and Problems

3.4.1. Determine the nature, the affine canonical equation, and the orthogonal canonical equation of the conics

$$\begin{aligned} &1)\ \ 3x^2 + 3y^2 - 2xy + 4x - 4y - 4 = 0, \\ &2)\ \ 3x^2 - y^2 + 2xy + 8x + 10y + 14 = 0, \\ &3)\ \ 9x^2 + 16y^2 + 24xy - 40x + 30y = 0. \end{aligned}$$

3.4.2. Determine the nature, the affine canonical equation, and the orthogonal canonical equation of the quadrics

$$1)\ \ 3x^2 + y^2 - z^2 + 6xz - 4y = 0,$$

$$2) \quad 2x^2 + y^2 + 3z^2 - 4yz + 2x - 6z + 1 = 0,$$

$$3) \quad x^2 + 2y^2 + 3z^2 + 2x - 4y - 12z + 8 = 0,$$

$$4) \quad 4x^2 - 9z^2 + 2xz - 8x - 4y + 36z - 32 = 0,$$

$$5) \quad 4x^2 + 2y^2 + z^2 - 4xy - 2yz - 2y + 2z - 4 = 0.$$

3.4.3. Discuss the nature of the quadrics

$$x^2 - y^2 + 2z^2 + 2\alpha xz - 1 = 0$$

for various values of $\alpha \in \mathbf{R}$.

3.4.4. Discuss the nature of the quadrics

$$y^2 + z^2 - 2\alpha xz + 12x - 5 = 0$$

for various values of $\alpha \in \mathbf{R}$.

3.4.5. Discuss the nature of the conics

$$x^2 + \alpha y^2 + 4xy - 6x + 2\beta y + 1 = 0$$

for various values of $\alpha, \beta \in \mathbf{R}$.

3.4.6. Consider the families of conics

$$i) \quad (\alpha - 1)x^2 + 2\beta xy - (\alpha + 1)^2 y + 2\alpha x + 2\beta y - (\alpha + 1) = 0,$$

$$ii) \quad \alpha x^2 + 2xy + \beta y^2 - 2\alpha x + \beta = 0.$$

Study the nature of the conics of each family for various values of $\alpha, \beta \in \mathbf{R}$ seen as orthogonal coordinates of a variable point in a plane.

Chapter 4

Geometric Transformations

4.1 Generalities

The theory of the previous chapters can be continued by the study of more complicated curves and surfaces, defined by more general *algebraic* and *transcendental* (i.e., beyond algebraic) equations. But, such developments are beyond the beginner's level, and they are discussed in advanced mathematical fields such as *Algebraic Geometry* and *Differential Geometry*. Here, we continue to discuss elementary subjects but, of a different nature. Namely, we will study elementary *geometric transformations* analytically.

Geometric objects, lines, planes, spaces, curves, surfaces etc., may be submitted to various changes. For instance, they may be moved around, compressed, enlarged etc. In order to study such changes, we look at the original object as a set of points $\mathcal{F} = \{A, B, \ldots\}$, and at the changed object as a second set of points $\mathcal{F}' = \{A', B', \ldots\}$, where the primes indicate the new *positions* of the original points. In this way, the change becomes a *correspondence* $\varphi : \mathcal{F} \to \mathcal{F}'$ between the two sets of points (geometric figures), and such correspondences are *geometric transformations.*

A very good illustration of the notions of a correspondence and a geometric transformation is *perspective* view of a painter which paints a flat object. Here we have the object $\mathcal{F}$ = the original set of points, and its image $\mathcal{F}'$ on a toil. Every point A of the object is joined to the point O where the eye of the painter is, and the obtained straight line cuts the plane of the toil at the corresponding point A' of the painting (see Fig. 4.1.1).

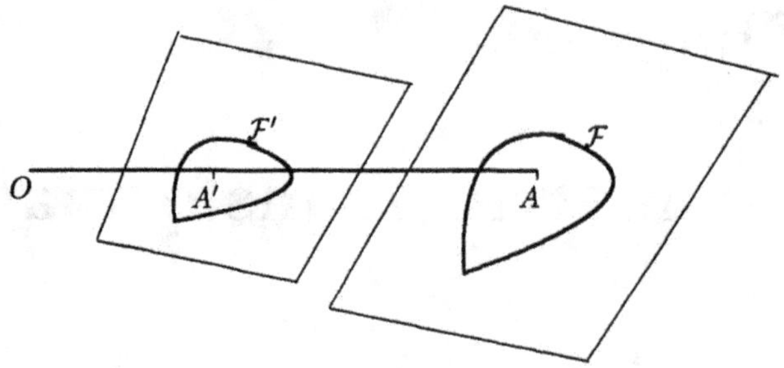

Fig. 4.1.1

Geometric transformations are important not only for geometry, but also for applications. In particular, *computer graphics* is a modern, practical discipline where geometric transformations are used extensively, because it represents real objects by images on the screen of a computer.

The formal definition of geometric transformations is given as follows

4.1.1 Definition. *A geometric transformation is a correspondence between two sets of geometric points,* $\varphi : \mathcal{F} \to \mathcal{F}'$*, such that: i)* $\forall A \in \mathcal{F}$*, one has a unique corresponding point* $A' := \varphi(A) \in \mathcal{F}'$*; ii)* $\forall A_1 \neq A_2 \in \mathcal{F}$*,* $A'_1 \neq A'_2$*; iii)* $\forall A' \in \mathcal{F}'$*, there exists* $A \in \mathcal{F}$ *such that* $A' = \varphi(A)$.

A is the *source* point, and A' is its *image*. Property ii) is known as the *injectivity* of the correspondence, and iii) as *surjectivity*. Together, ii) and iii) mean the *bijectivity* of φ i.e., the correspondence is *one-to-one* in both directions: every source has a unique image and vice versa. The set $\mathcal{F}$ is the *domain*, and $\mathcal{F}'$ is the *range* of φ. Correspondences which are not one-to-one are also very important, but we will not discuss them here.

We need a few more important, general facts about correspondences.

4.1.2 Definition. *i) Let* $\varphi : \mathcal{F} \to \mathcal{F}'$ *be a geometric transforma-*

tion, and consider the transformation $\varphi^{-1} : \mathcal{F}' \to \mathcal{F}$ *defined by* $A = \varphi^{-1}(A') \iff A' = \varphi(A)$; φ^{-1} *exists because* φ *is bijective, and it is called the inverse of* φ. *ii) Let* $\varphi : \mathcal{F} \to \mathcal{F}'$, *and* $\psi : \mathcal{F}' \to \mathcal{F}''$ *be two geometric transformations, and define the transformation* $\psi \circ \varphi : \mathcal{F} \to \mathcal{F}'$ *as the successive application of the correspondences* φ *and then* ψ *(in this order! i.e.,* $\psi \circ \varphi(A) := \psi(\varphi(A)))$; *then,* $\psi \circ \varphi$ *is called the composed transformation of the given ones, and composition may be seen as an algebraic operation with correspondences. iii) For any set* $\mathcal{F}$, *there exists a transformation which leaves every point unchanged; we denote it* $id_{\mathcal{F}}$, *and call it the identical transformation of* $\mathcal{F}$. *(Thus,* $id_{\mathcal{F}}(A) = A$, $\forall A \in \mathcal{F}$.*)*

The last notion makes sense since, generally, we are allowed to transform a figure into itself by interchanging in a precised way the *places* of its points.

4.1.3 Proposition. *The composition of geometric transformations has the following properties: i) for any transformations, if one of the two sides of the equality*

$$\varphi_3 \circ (\varphi_2 \circ \varphi_1) = (\varphi_3 \circ \varphi_2) \circ \varphi_1,$$

exists, the other side also exists, and the equality holds (this is the associativity of composition); ii) for every geometric transformation $\varphi : \mathcal{F} \to \mathcal{F}'$, *we have*

$$\varphi^{-1} \circ \varphi = id_{\mathcal{F}}, \ \varphi \circ \varphi^{-1} = id_{\mathcal{F}'}.$$

Proof. Everything is straightforward. In particular, associativity holds because the two sides of the equality of i) mean the same thing: apply the correspondences successively in the indicated order. As a consequence, composition may be extended to any finite number of transformations, if the range of each of them is contained in the domain of the next. Q.E.D.

4.1.4 Remark. *Generally, composition of transformations is not commutative, i.e., even if both exist,* $\varphi \circ \psi \neq \psi \circ \varphi$. *We will have plenty of examples of transformations, and noncommutativity in the remainder of this chapter.*

4.1.5 Definition. *Let $\mathcal{F}$ be a set of points. A nonvoid family $G = \{\varphi : \mathcal{F} \to \mathcal{F}\}$ of geometric transformations of $\mathcal{F}$ is called a group of transformations or a transformation group if i) $\forall \varphi, \psi \in G$, $\psi \circ \varphi$ also belongs to G, and ii) $\forall \varphi \in G$, the inverse φ^{-1} belongs to G.*

We notice that this definition is consistent with the definition of a group (see Remark 1.2.5). In the present case, the operation is the composition of transformations, and, for a fixed $\mathcal{F}$, it always exists. Then, $id_{\mathcal{F}} = \varphi^{-1} \circ \varphi \in G$ because of i) and ii) above, and $\forall \varphi \in G$, $\varphi \circ id_{\mathcal{F}} = id_{\mathcal{F}} \circ \varphi = \varphi$. This corresponds to property b) of a group. Finally, the existence of φ^{-1} corresponds to property c) of the definition of a group (Remark 1.2.5). The family of *all the transformations* $\mathcal{F} \longrightarrow \mathcal{F}$ obviously is a group of transformations. If G_1 and G_2 are groups of transformations of the same $\mathcal{F}$, and if G_1 is contained in G_2, we say that G_1 is a *subgroup* of G_2.

The notion of a transformation group is of fundamental importance in mathematics and in physics. In particular, *the study of the properties of a geometric figure (plane, space, etc.) which are invariant by a fixed group of transformations G is called the G-geometry of that figure.* This definition of geometry is due to Felix Klein, and is contained in a famous mathematical document: *the Erlangen Program*, the inaugural dissertation of Klein at the University of Erlangen (Germany), in 1872.

Exercises and Problems

4.1.1. Let $\varphi, \psi : \mathcal{F} \longrightarrow \mathcal{F}$ be two geometric transformations. Prove that $(\psi \circ \varphi)^{-1} = \varphi^{-1} \circ \psi^{-1}$.

4.1.2. Does the set of all the geometric transformations of all the different geometric figures of space form a transformation group?

4.2 Affine Transformations

The importance of affine and orthogonal frames suggests that it should be equally important to define

4.2.1 Definition. *Let $\mathcal{F}, \mathcal{F}'$ be the sets of points of either two straight lines d, d' or two planes α, α' or of space, considered twice (the cases $d =$*

*d', $\alpha = \alpha'$ are possible). Then, a geometric transformation $\varphi : \mathcal{F} \to \mathcal{F}'$ is an **affine transformation**, respectively an **orthogonal transformation**, if there exists a pair of affine (orthogonal) frames in $\mathcal{F}$, $\mathcal{F}'$, respectively, such that $\forall M \in \mathcal{F}$ the coordinates of M, and of $M' := \varphi(M)$ with respect to these two frames are equal.*

In other words, if we agree that two frames as in Definition 4.2.1 are *corresponding* one to the other by φ, the transformation is affine (orthogonal) if there are corresponding affine (orthogonal) frames such that the *equations of* φ with respect to these frames are

$$x' = x, \; y' = y, \; z' = z. \tag{4.2.1}$$

Of course, in (4.2.1), (x, y, z) are the coordinates of $M \in \mathcal{F}$, and (x', y', z') are the coordinates of the corresponding point $M' \in \mathcal{F}'$, and the equations are written for space. For straight lines, we only have x, and, for planes, we only have (x, y).

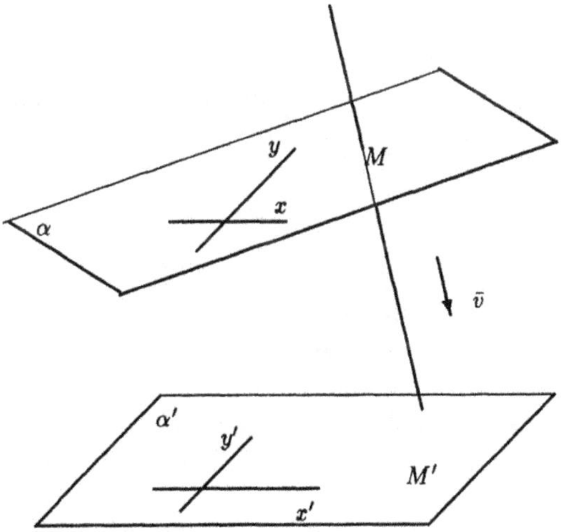

Fig. 4.2.1

4.2.2 Example. Let α and α' be two arbitrary, fixed planes, and let

$\bar{v} \neq \bar{0}$ define an arbitrary, fixed direction which is not parallel to any of the two planes. Define $\varphi : \alpha \to \alpha'$ by sending each point $M \in \alpha$ to $M' := \varphi(M) = \alpha' \cap d(M, \bar{v})$, where $d(M, \bar{v})$ is the straight line through M, with direction $\bar{v}$ (see Fig. 4.2.1).

This transformation is called a *parallel projection* or an *affine perspective* affine perspective between the two planes. It is obvious that, if we take any affine frame of α, and project it to α', parallel to $\bar{v}$, we obtain a corresponding frame of α', such that (4.2.1) holds for x and y. The reader will check that this is indeed true by noticing that parallel projection preserves simple ratios, and recalling that the affine coordinates are simple ratios (Proposition 1.3.7). Hence, the transformation defined above is affine. But, generally, this is not an orthogonal transformation, since the projection of an orthogonal frame may not be orthogonal. This last remark shows that, generally, an affine transformation may change the lengths and angles of the geometric figures. Thus, while every orthogonal transformation is affine (why?), an affine transformation may not be orthogonal.

4.2.3 Proposition. *A transformation $\varphi : \mathcal{F} \to \mathcal{F}'$ between two lines, (planes, spaces) is affine (orthogonal) iff the equations of φ with respect to arbitrary, affine (orthogonal) frames look like a change (1.3.10) of affine (orthogonal) coordinates i.e.,*

$$\tilde{x}'_j = \sum_i b_{ij}\tilde{x}_i + \beta_j. \tag{4.2.2}$$

Proof. In (4.2.2), $(\tilde{x}_i)$ are the coordinates of $M \in \mathcal{F}$, and $(\tilde{x}'_i)$ are the coordinates of the corresponding point $M' \in \mathcal{F}'$, with respect to arbitrary frames, and the equations are such that they may be solved for $\tilde{x}_i$. The summation index runs from 1 to 1,2 and 3, corresponding to the case of lines, planes and space, respectively.

If φ is affine, there are corresponding affine coordinates, with the equations (4.2.1), and $\tilde{x}'_i$ are given by an affine change of the form (1.3.10) of the coordinates x'_i. But, also the x_i are given by an affine change of the coordinates $\tilde{x}_i$. Accordingly, (4.2.1) becomes a formula of the type (4.2.2).

Conversely, if φ is given by equations (4.2.2), and if we define new affine coordinates of $\mathcal{F}$ by

$$\hat{x}_j = \sum_i b_{ij}\tilde{x}_i + \beta_j,$$

the new equations of φ will be $\tilde{x}'_j = \hat{x}_j$, and we see that we have an affine transformation.

The orthogonal case is settled similarly, only that the matrix (b_{ij}) of the coefficients of equation (4.2.2) is *orthogonal* (see the meaning of this term in Section 1.4.). Q.E.D.

Another important result is

4.2.4 Proposition. *For any line, plane or space $\mathcal{F}$, the set of all the affine (orthogonal) transformations from $\mathcal{F}$ to itself is a group of transformations.*

Proof. From equations (4.2.1), it follows straightforwardly that composed and inverses of affine (orthogonal) transformations are affine (orthogonal). Q.E.D.

In the affine case, the group is called the *affine group* of $\mathcal{F}$, and is denoted by $A(\mathcal{F})$. Similarly, we have the *orthogonal group* $O(\mathcal{F})$. Since the affine (orthogonal) properties of geometric figures were exactly the properties invariant by affine (orthogonal) changes of coordinates, a comparison with Proposition 4.2.3 shows that these are the properties which remain unchanged by affine (orthogonal) transformations. This means that affine geometry is the geometry of the affine groups, and orthogonal (Euclidean) geometry is the geometry of the orthogonal groups, in the sense of the Erlangen Program mentioned at the end of section 4.1.

4.2.5 Proposition. *If $\varphi : \mathcal{F} \to \mathcal{F}'$ is an affine transformation between lines, planes or in space, and if $\mathcal{V}(\mathcal{F})$, $\mathcal{V}(\mathcal{F}')$ are the linear spaces (Remark 1.2.12) of the vectors of these figures, φ induces a one-to-one correspondence $l : \mathcal{V}(\mathcal{F}) \to \mathcal{V}(\mathcal{F}')$ with the following properties:*
i) $l(\bar{v}_1 + \bar{v}_2) = l(\bar{v}_1) + l(\bar{v}_2)$; ii) $l(\alpha\bar{v}) = \alpha l(\bar{v})$.

Proof. Take $\bar{v} = \vec{AB}$, and define $l(\bar{v}) = \vec{A'B'}$, where $A' = \varphi(A)$, $B' = \varphi(B)$. Then, with respect to corresponding affine frames, $\bar{v}$ and $l(\bar{v})$ have equal coordinates. This shows that l does not depend on the choice of the representative segment $\vec{AB}$ of $\bar{v}$, and that i) and ii) hold. Q.E.D.

l is called the *linear transformation associated with* φ. Since the coordinates of $\vec{AB}$ are the differences of the coordinates of B and A, we see that, with respect to arbitrary frames, if φ has equations (4.2.2), l has the equations

$$\tilde{v}'_j = \sum_i b_{ij}\tilde{v}_i, \tag{4.2.3}$$

which is exactly the homogeneous, linear part of the equations of φ.

The following is a collection of important properties of affine transformations

4.2.6 Proposition. *i) An affine transformation sends three collinear (noncollinear) points to three collinear (noncollinear) points, and it preserves the simple ratio of the points. ii) An affine transformation sends four coplanar (noncoplanar) points to four coplanar (noncoplanar) points. iii) An affine transformation sends straight lines and planes to straight lines and planes, respectively, and it preserves parallelism.*

Proof. If we use corresponding frames, points which correspond each to the other by the transformation have equal coordinates, and everything mentioned in proposition has the same coordinate expression before and after the transformation. Q.E.D.

In fact, it is easy to prove

4.2.7 Proposition. *A geometric transformation* φ *between lines, planes, or in space is affine iff it preserves collinearity, noncollinearity, coplanarity, noncoplanarity and the simple ratios of points.*

Proof. If φ is affine, we already know that these properties are preserved. Conversely, from the above preservation hypotheses, it follows that φ sends straight lines to straight lines, planes to planes, and affine frames to affine frames. Then, since the affine coordinates are simple ratios (Proposition 1.3.7), it follows that our transformation has equations (4.2.1) with respect to the corresponding frames. Q.E.D.

4.2.8 Remark. *As a matter of fact, it suffices to ask the transformation to preserve collinearity only, in the case of planes and space, and add the preservation of the simple ratios in the case of straight lines, in order to deduce that the transformation is affine. But the proof is much longer.*

The following result is known as the *fundamental theorem for affine transformations*

4.2.9 Theorem. *i) An affine transformation is uniquely defined by a corresponding pair of affine frames. ii) An affine transformation is uniquely defined by the following configurations: 1) two corresponding pairs of distinct points, in the case of straight lines; 2) two corresponding triples of noncollinear points, in the case of planes; 3) two corresponding quadruples of noncoplanar points, in space.*

Proof. For i), we just have to use equations (4.2.1). For ii), if we take the case of space, for instance, and if the corresponding quadruples are $(ABCD)$, $(A'B'C'D')$, we will take the frames with origins A, A', and bases $(\vec{AB}, \vec{AC}, \vec{AD})$, $(\vec{A'B'}, \vec{A'C'}, \vec{A'D'})$, respectively, as corresponding frames, and get the required transformation as in case i). The cases of lines and planes are to be treated similarly. Q.E.D.

Accordingly, the usual way of establishing affine transformations is by giving the corresponding configurations as in Theorem 4.2.9, and by determining the coefficients of equations (4.2.2) which ensure the correspondence of the given configurations.

4.2.10 Example. Find a plane affine transformation which sends the vertices of a triangle ΔABC to their symmetric points with respect to the midpoints of the opposite sides.

Solution. Let us choose the affine frame of the plane such that $A(0,0)$, $B(1,0), C(0,1)$. Then, the corresponding symmetric points of the problem will be $A'(1,1)$, $B'(-1,1)$, $C'(1,-1)$. Assume that the required affine transformation φ is given by

$$x' = a_{11}x + a_{12}y + a_{10}, \quad y' = a_{21}x + a_{22}y + a_{20}.$$

Then, $\varphi(0,0) = (1,1)$ implies $a_{10} = a_{20} = 1$, $\varphi(1,0) = (-1,1)$ implies $a_{11} = -2$, $a_{21} = 0$, $\varphi(0,1) = (1,-1)$ implies $a_{12} = 0$, $a_{22} = -2$. Hence, with respect to the chosen frame, the equations of φ are

$$x' = -2x + 1, \quad y' = -2y + 1.$$

We end by indicating some important particular types of affine transformations. First, a geometric transformation on a straight line, in a plane, or in space is called a *translation* if $\forall A, B$, we have

$$\vec{AA'} = \vec{BB'}. \tag{4.2.4}$$

To fix the ideas, let us look at translations in space. They are necessarily affine transformations since, with respect to a fixed frame, we must have the equations

$$x_i(A') = x_i(A) + v_i, \tag{4.2.5}$$

where $\bar{v}(v_i)$ is the *constant* vector given by (4.2.4), called *the vector of the translation.* Clearly, an affine transformation is a translation iff the linear transformation of vectors associated with it is identical. It is also clear that two frames are corresponding to one another by a translation iff one is obtained from the other by a usual parallel translation in space. Thus, our present translations are just the well-known parallel translations encountered in Chapter 1. Moreover, this remark also shows that, in fact, the translations are orthogonal transformations. Furthermore, we ask the reader to check by himself that the set of all the translations of space is a commutative transformation group T where the composition of translations is given by the addition of the vectors of the translations. Similar results hold for translations in a fixed plane or along a fixed straight line.

A second type of transformations which we want to define is *homothety.* We will look at homotheties in space but, it is easy to understand that similar transformations exist on a fixed plane or straight line. The *homothety* of *center* O and *ratio* $k \neq 0$, $k \in \mathbf{R}$, is the geometric transformation which leaves the point O on the spot, and sends any other point M to the point M' which belongs to the straight line OM, and satisfies the condition

$$\vec{OM'} = k\vec{OM}. \tag{4.2.6}$$

With respect to a fixed frame of origin O, the equations of the homothety are

$$x_i(M') = kx_i(M), \tag{4.2.7}$$

and this shows that a homothety is an affine transformation, necessarily. (But, if $|k| \neq 1$, it is not an orthogonal transformation since the unit length of the basic vectors of an orthonormal frame is not preserved.) We ask the reader to check that the homotheties of space, with the same center, make a commutative transformation group, where the composition of homotheties is just the multiplication of their ratios.

The example of homotheties shows the importance of *fixed points* of a geometric transformation i.e., points which the transformation doesn't change. Similarly, we may look at *fixed lines*, *fixed planes*, etc., i.e., lines (planes) whose points, while not fixed, are sent to points of the same line (plane).

We do not intend to study such fixed objects here. Let us only give one more definition. An affine transformation which has a fixed point O, is called a *centroaffine transformation* with *center* O. The family of all the centroaffine transformations (of space, say) with the same center O is a noncommutative transformation subgroup of the affine group. Moreover, any affine transformation φ is decomposable as a composition of a centroaffine transformation ψ of center O and a translation. This is an immediate consequence of formulas (4.2.2); the homogeneous, linear part is the centroaffine component (we use a frame with origin at the desired center), and the addition of β_j is the translation.

In order to define some more interesting subgroups of the affine group, let us first look at the affine transformation (4.2.2), and let us write its associated, linear transformation of vectors using a matrix notation, say

$$v' = Cv, \tag{4.2.8}$$

where v, v' are the columns of the coordinates $\bar{v}$ and of its image $\bar{v}'$, and C is the matrix of the transformation. Let us assume that the coordinates used in (4.2.8) are orthogonal coordinates. Then, (4.2.8) shows that a mixed product changes by the following rule

$$(\bar{v}'_1, \bar{v}'_2, \bar{v}'_3) = det(C)(\bar{v}_1, \bar{v}_2, \bar{v}_3),$$

and we see that the transformation preserves *oriented volumes* (see Proposition 1.4.19) iff $det(C) = 1$. If in (4.2.8) the coordinates are not orthogonal, we go over to an orthogonal frame by $v = S\tilde{v}$, $v' = S\tilde{v}'$, and, for the new equation $\tilde{v}' = \tilde{C}\tilde{v}$, we will have $\tilde{C} = S^{-1}CS$. This shows that $det(C) = det(\tilde{C})$. Thus, the preservation of oriented volumes is characterized by $det(C) = 1$ with respect to any affine frame. Similarly, working with a vector product, in plane, the condition $det(C) = 1$ characterizes the affine transformations which preserve the oriented area of a parallelogram. Obviously, the weaker condition $det(C) > 0$ only means that the transformation is *orientation preserving.*

It follows that the affine groups have the subgroup of orientation preserving transformations, called *direct transformations.* The transformations which reverse the orientation, called *indirect transformations,* are characterized by $det(C) < 0$. (The latter transformations do not form a subgroup. Why?) Then, there also is the subgroup of *equiaffine transformations* of space (plane), which, by definition, consists of the affine transformations which preserve the oriented volume of a parallelepiped (the oriented area of a parallelogram, respectively), and are characterized by $det(C) = 1$.

Exercises and Problems

4.2.1. Prove that the translations of space form a commutative group of transformations, where composition corresponds to the addition of the translation vectors.

4.2.2. Prove that the homotheties of a given center O form a commutative group of transformations, where composition corresponds to the multiplication of the homothety ratios.

4.2.3. Under what condition are two trapeziums affinely equivalent in the plane? (The affine equivalence of two figures means the existence of an affine transformation which sends one figure onto the other figure.)

4.2.4. Find all the affine transformations of a plane which keep fixed the hyperbola $xy = c$.

4.2.5. An affine transformation φ of a plane α is called an *affine perspective* if it has a line of fixed points (called *axis*), and, $\forall A \in \alpha$ and with the notation $A' = \varphi(A)$, the lines AA' are parallel. i) Find the equation of an affine perspective with respect to an affine frame which has Ox as its line of fixed points, and Oy in the direction of AA'.
ii) Is the transformation

$$x' = 4x + y - 5, \;\; y' = 6x + 3y - 10$$

an affine perspective?

4.2.6. Prove that for any plane affine transformation there exists a (possibly nonunique) pair of orthogonal unmarked directions whose orthogonality is preserved by the transformation. (These are called the *principal directions* of the transformation. Hint: Transform a circle into an ellipse and show that the axes of the ellipse are images of perpendicular diameters of the circle by the transformation.)

4.2.7. Find the principal directions of the affine transformation φ which has the equations

$$x' = 3x - 6y + 5, \quad y' = 2x + 3y - 7$$

with respect to an orthogonal frame.

4.2.8. Prove that if the sides of a parallelogram Π are tangent to an ellipse Γ the diagonals of Π are conjugated diameters of Γ. And, in particular, if Π is a rhombus its diagonals are the symmetry axes of Γ. (Hint: Apply an affine transformation which transforms Γ into a circle.)

4.2.9. Using an orthogonal frame, find all the affine transformations of space which send every vector to a vector orthogonal to the source vector.

4.3 Orthogonal Transformations

We know that the orthogonal transformations are affine. Therefore, they share with all the affine transformations the properties contained in Propositions 4.2.4, 4.2.6, but we have new properties too.

4.3.1 Proposition. *An orthogonal transformation preserves the scalar product of two vectors, and, consequently, it preserves the length of a vector, the distance between two points, the angle of two directions, and nonoriented areas and volumes.*

Proof. With respect to an orthogonal frame, the scalar product is expressed by formula (1.4.7) which contains only the coordinates of the vectors. Hence, if we express our transformation by the equations (4.2.1), we are done. Q.E.D.

As a matter of fact, one can prove that a geometric transformation between lines, planes, or in space is orthogonal if and only if it is distance preserving. Indeed, it is obvious that, if a transformation preserves the distance between two arbitrary points, it will send collinear points to collinear points, and it will preserve the simple ratios. Hence, as we saw in the previous section, the transformation is affine, and it sends an orthogonal frame to an affine frame. Moreover, since the lengths are preserved so are the angles (why?), and the image frame of an orthogonal frame must also be an orthogonal frame. Because of this result, the orthogonal transformations are also called *isometries*. If, moreover, the transformation is orientation preserving, it is called a *direct isometry* or, if it is of a line, plane or space to itself, a *motion*. From Proposition 4.3.1 and the explanations given at the end of the previous section, we see that a direct orthogonal transformation is equiaffine, and if it has equation (4.2.8) for vectors, it satisfies $det(C) = 1$. The motions of space (line, plane) define a group of transformations. The indirect isometries are characterized by $det(C) = -1$.

For orthogonal transformations, the fundamental Theorem 4.2.9 must be changed as follows

4.3.2 Theorem. *i) An orthogonal transformation is uniquely defined by a pair of corresponding orthogonal frames. ii) An orthogonal transformation is uniquely defined by the following configurations: 1) two corresponding, oriented segments of equal, nonzero length, in the case of straight lines; 2) two corresponding, congruent triangles, in the case of planes; 3) two corresponding, congruent tetrahedra, in space.*

Proof. We recall that the word *congruence* is a synonymous of equality in the case of nonoriented triangles, tetrahedra (and geometric figures, generally), and equality refers to all the corresponding sides, edges, angles and dyhedral angles. The arguments of the proof are the same as for Theorem 4.2.9, but, for ii) we have to justify that the given configurations lead to pairs of corresponding orthogonal (not necessarily positive) frames. This is easy. For instance, if $\triangle ABC = \triangle A'B'C'$, we may take A, A' as corresponding origins, the vectors $\bar{\imath} := \vec{AB}/|\vec{AB}|$, $\bar{\imath}' := \vec{A'B'}/|\vec{A'B'}|$ as the first vectors of the corresponding bases, and $\bar{\jmath} \perp \bar{\imath}, \bar{\jmath}' \perp \bar{\imath}'$ of unit length and such that the orientation of $\widehat{(\bar{\imath}, \bar{\jmath})}$, $\widehat{(\bar{\imath}', \bar{\jmath}')}$ are the same as the orientation of $\widehat{BAC}$ and $\widehat{B'A'C'}$, respectively. The reader will give similar constructions on a line and in space. Q.E.D.

Note that the transformation provided by Theorem 4.3.2 is direct iff the described corresponding configurations have the same orientation, and it is indirect if the orientations are opposite.

Now, we will study some important, particular, orthogonal transformations, and give some results about the structure of the general isometries.

First, we remember that the translations are orthogonal transformations, motions, in fact (why?).

Then, a plane motion with a fixed point, called *center*, is called a *rotation* around that center. It is easy to obtain the equations of a rotation. Namely, we have

4.3.3 Proposition. *If O is the center of a plane rotation, the angle $\theta := \widehat{AOA'}$, where A is an arbitrary point and A' is the rotated point, is a constant angle (independent of A), called the rotation angle, and, with respect to any orthogonal system of coordinates (x, y) with origin at O, the equations of the rotation are*

$$\tilde{x} = x\cos\theta - y\sin\theta, \quad \tilde{y} = x\sin\theta + y\cos\theta, \tag{4.3.1}$$

where $A(x, y)$, and $A'(\tilde{x}, \tilde{y})$.

Proof. If $(\bar{\imath}, \bar{\jmath})$ is an orthonormal basis, the frame $(O, \bar{\imath}, \bar{\jmath})$ has a corresponding orthogonal frame $(O, \bar{\imath}', \bar{\jmath}')$, where (see (1.4.14), and Fig. 4.3.1)

$$\bar{\imath}' = \bar{\imath}\cos\varphi + \bar{\jmath}\sin\varphi, \quad \bar{\jmath}' = -\bar{\imath}\sin\varphi + \bar{\jmath}\cos\varphi, \tag{4.3.2}$$

where $\varphi := \widehat{\bar{\imath}, \bar{\imath}'}$.

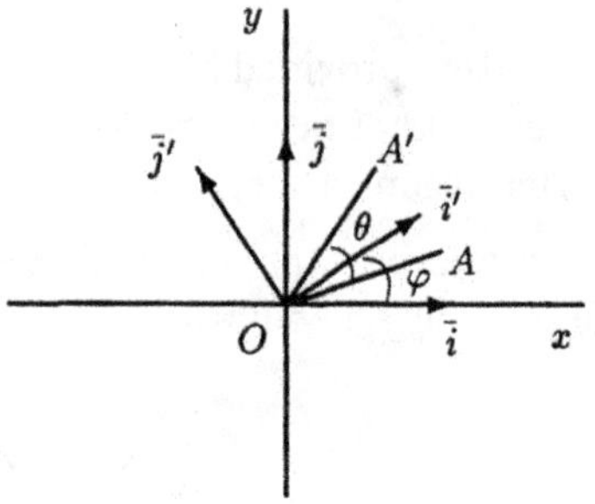

Fig. 4.3.1

Since the coordinates of corresponding points with respect to the corresponding frames are equal, we have

$$\vec{OA} = x\bar{i} + y\bar{j},$$

and

$$\vec{OA'} = x\bar{i}' + y\bar{j}' = \bar{i}(x\cos\varphi - y\sin\varphi) + \bar{j}(x\sin\varphi + y\cos\varphi). \tag{4.3.3}$$

(4.3.3) shows that the coordinates of A' with respect to the original frame are given by (4.3.1), except for the fact that the angle is φ. But, using (4.3.3) again, it follows easily that $\widehat{AOA'} = \varphi$ (compute $\cos\widehat{AOA'}$, and $\sin\widehat{AOA'}$). Q.E.D.

Furthermore, if an orthogonal transformation has two fixed points P, Q, every point of the line PQ is fixed (use Theorem 4.3.2). A space motion which has such a line of fixed points is called a *rotation (revolution)* around the fixed line, called the *axis.* If we take an orthogonal frame with the rotation axis as Oz, the rotation induces a rotation around O in the xy-plane (why?), and Proposition 4.3.3 yields

4.3.4 Proposition. *If d is the axis of a rotation, the angle θ between the planes (d, A) and (d, A'), where A is an arbitrary point and A' is the rotated point, is a constant angle (independent of A), called the rotation angle, and, with respect to any orthogonal system of coordinates (x, y, z) with the z-axis along d, the equations of the rotation are*

$$\tilde{x} = x\cos\theta - y\sin\theta,\quad \tilde{y} = x\sin\theta + y\cos\theta,\quad \tilde{z} = z, \tag{4.3.4}$$

where $A(x,y,z)$, and $A'(\tilde{x},\tilde{y},\tilde{z})$.

Proof. The only novelty is the last formula (4.3.4). But this is clear, since an orthogonal transformation preserves lengths and angles. Q.E.D.

A space motion which is the composition of a rotation around an axis and a translation by a vector parallel to this axis is called a *helicoidal transformation*. If we take the rotation axis as z-axis, the equations of the helicoidal transformation will obviously be

$$\tilde{x} = x\cos\theta - y\sin\theta,\quad \tilde{y} = x\sin\theta + y\cos\theta,\quad \tilde{z} = z + h, \tag{4.3.5}$$

where h is a constant number.

A space motion which has a fixed point O, called *center*, is called a *space rotation* around that center. However, as a matter of fact we have

4.3.5 Proposition. *In space, a rotation around a point always has an axis, and it is a rotation around this axis.*

Proof. A straight line through the fixed point, say O, with direction $\bar{v} \neq \bar{0}$ is a rotation axis iff the associated linear transformation sends $\bar{v}$ to a proportional vector $\lambda\bar{v}$. If we look for such a vector $\bar{v}$ by using for the transformation equations (4.2.3), we see that the coordinates v_i of $\bar{v}$ must be a nontrivial solution of a *characteristic system*

$$\sum_{i=1}^{3}(b_{ji} - \lambda\delta_{ji})v_i = 0,$$

which looks like the system (3.3.37) associated with a quadric. Hence, first, λ must be a real solution of the equation which corresponds to (3.3.39) i.e.,

$$det(b_{ji} - \lambda\delta_{ji}) = 0.$$

Since this equation is of the third degree, and it has real coefficients, it must have a real solution, and we are done. Q.E.D.

4.3.6 Remark. *On a straight line, the only motion which has a fixed point is the identical transformation. (Why?)*

By the same argument as for affine transformations (i.e., by looking at equations (4.2.2)), it follows that any orthogonal transformation is the composition of a transformation with a fixed point and a translation. In particular, any motion is the composition of a rotation and a translation.

4.3.7 Definition. *A geometric transformation σ of a straight line, plane or space is a symmetry or a reflection with respect to a point (center), or a line axis or a plane if the points A and $A' := \sigma(A)$ are symmetric with respect to the point, line or plane in the usual sense i.e., they are on different sides and at equal distance. (See Fig. 4.3.2.)*

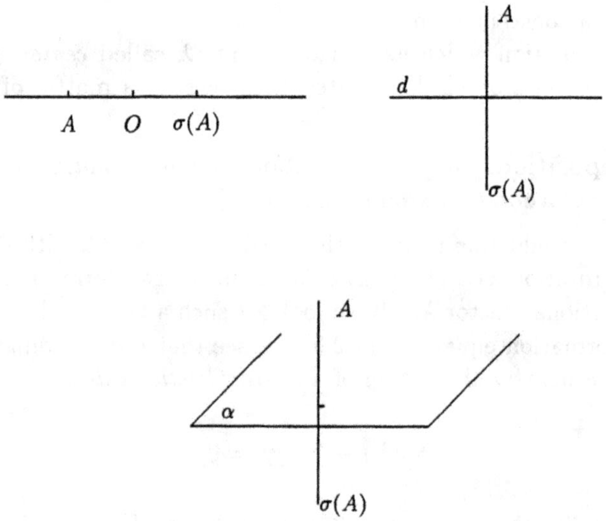

Fig. 4.3.2

For instance, on the x-axis, the symmetry with respect to the origin has the equation $x' = -x$, in the (x, y)-plane, the symmetry with respect to the x-axis has the equations $x' = x$, $y' = -y$, and, in space, the symmetry with respect to the xy-plane has the equations $x' = x$, $y' = y$, $z' = -z$.

It is obvious from the definition of a symmetry that it preserves distance. This fact, and Proposition 4.2.7 show that a symmetry is an orthogonal transformation. Moreover, the equations of symmetries written above show that, on a line (plane, space), a symmetry with respect to a point (line, plane, respectively) is an indirect transformation (its determinant is -1). We also notice the interesting facts that a symmetry σ fixes its center, and every point of its axis or plane, respectively, and it satisfies the equality $\sigma^2 := \sigma \circ \sigma = id.$, or, equivalently, $\sigma = \sigma^{-1}$.

4.3.8 Proposition. *On a line (plane, space), any indirect isometry is the composition of a motion and a symmetry with respect to a point (line, plane). More exactly: i) On a line, an indirect orthogonal transformation with a fixed point (center) is a symmetry. ii) In plane, an indirect orthogonal transformation with a fixed point (center) is the composition of a rotation around this point and a symmetry with respect to a line through the center. iii) In space, an indirect orthogonal transformation with a fixed point (center) is the composition of a rotation with a symmetry with respect to a plane through the center.*

Proof. If the isometry is τ, and σ is an arbitrary symmetry, $\tau' := \tau \circ \sigma$ is a motion, and $\tau = \tau' \circ \sigma$, as required.

i) is rather obvious. For ii), if the transformation is again τ, let σ be the symmetry with respect to any line through the center O, and $\tau' = \tau \circ \sigma$. As the composition of two indirect transformations, τ' is direct, and it has the fixed point O; hence, τ' is a rotation, and $\tau = \tau' \circ \sigma$ as required (use $\sigma^{-1} = \sigma$). The proof of iii) is completely similar. Q.E.D.

Now, we will prove a theorem which shows the fundamental role of symmetries.

4.3.9 Theorem. *Every orthogonal transformation of space (plane, straight line) is the composition of a certain number of reflections with respect to planes (or lines or points, respectively).*

Proof. The proof of the spatial case includes the other two cases, as we will see later. In view of Propositions 4.3.8, 4.3.5, it is enough to prove the assertion of the theorem for rotations around an axis and for translations only.

First, we notice that the symmetry σ with respect to a plane $z = c$ has the equations

$$x' = x, \quad y' = y, \quad z' - c = c - z,$$

i.e.,

$$x' = x, \quad y' = y, \quad z' = -z + 2c \tag{4.3.6}$$

(see Fig. 4.3.3).

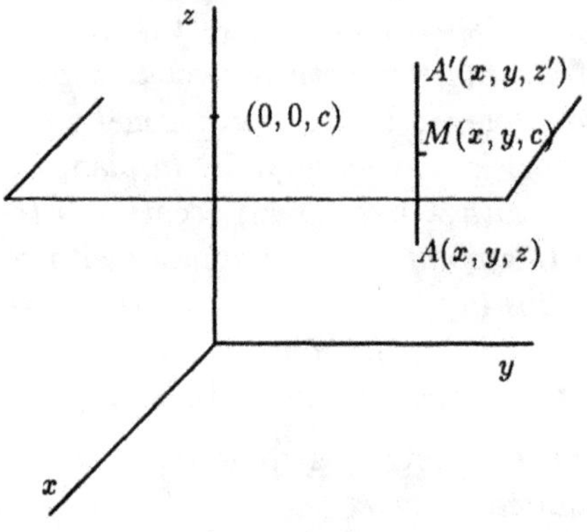

Fig. 4.3.3

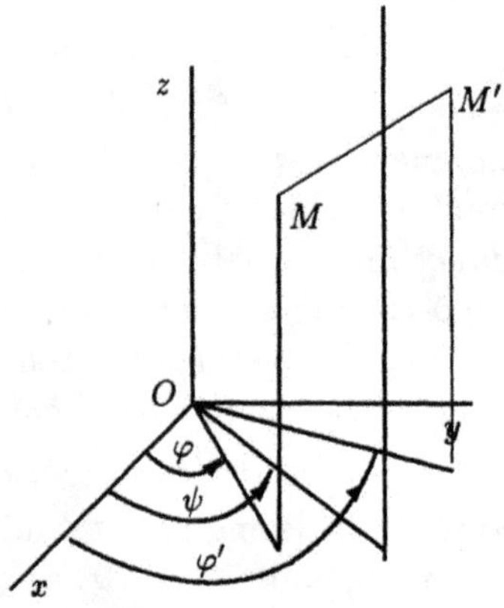

Fig. 4.3.4

Now, if we have a translation τ, and if we choose the z-axis in the direction of τ, the translation has equations

$$\tilde{x} = x, \quad \tilde{y} = y, \quad \tilde{z} = z + a,$$

whence it follows that we may see it as the composition of the transformations

$$\tilde{x} = x', \quad \tilde{y} = y', \quad \tilde{z} = -z', \tag{σ_1}$$

$$x' = x, \quad y' = y, \quad z' = -z - 2\frac{a}{2}, \tag{σ_2}$$

which are the symmetries with respect to the planes $z = 0$ and $z = -a/2$, respectively.

To get the proof of the same fact in plane or on a line, we use the (y, z)-part, or the z-part of the previous equations only.

Now, in order to decompose a rotation of angle θ around, say, the z-axis, we will use the cylindrical coordinates (ρ, φ, z) defined in section 3.1 (Fig. 3.1.4). Then, the rotation has the equations

$$\rho' = \rho, \quad z' = z, \quad \varphi' = \varphi + \theta, \tag{4.3.7}$$

which look like the equations of a translation.

Similarly, the equations of a symmetry with respect to a plane through the z-axis are of the form (check!)

$$\rho' = \rho, \quad z' = z, \quad \varphi' = -\varphi + 2\psi, \tag{4.3.8}$$

where ψ is the angle between the xz-plane and the plane of the reflection (see Fig. 4.3.4). And, (4.3.8) are similar to (4.3.6).

Now, we may write the rotation as the composition of two symmetries σ_1, σ_2 with respect to planes which pass through Oz, and which have the same equations as for a translation but, in cylindrical coordinates.

In the case of a plane rotation, we use polar coordinates (section 3.1). In the case of a line, there are no nonidentical rotations. In any case, the identical transformation is $id = \sigma \circ \sigma$ for any symmetry σ. Q.E.D.

An interesting class of affine transformations, which we will not study in detail is given by

4.3.10 Definition. *A transformation which is the composition of an orthogonal transformation and a homothety is called a similarity.*

Obviously, a similarity preserves the measure of an angle, but it multiplies a distance by the ratio of the corresponding homothety. The similarities form a group of transformations which is an important subgroup of the affine group. In some books, it is the geometry of this subgroup which is called Euclidean geometry (in the sense of the Erlangen Program).

We end this section and chapter by mentioning some very interesting geometric transformations which are neither orthogonal nor affine.

4.3.11 Definition. *Let O be a point in plane (space), and let $\mathcal{M}$ be the set of all the points, except O. The inversion of pole O, and power $k \in \mathbf{R}$ is the geometric transformation of $\mathcal{M}$ which sends $M \in \mathcal{M}$ to $M' \in \mathcal{M}$*

which is collinear with O and M, and satisfies $a(\vec{OM})a(\vec{OM'}) = k$ *(see Fig. 4.3.5).*

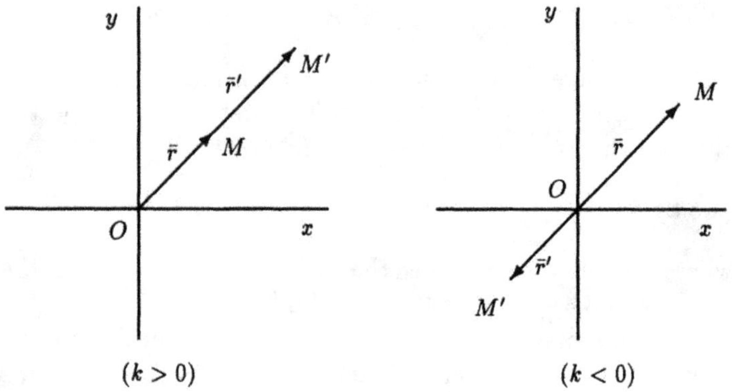

Fig. 4.3.5

If we use an orthogonal frame with origin O, it follows that the inversion has the equation

(4.3.9) $$\bar{r}' = \frac{k\bar{r}}{|\bar{r}|^2},$$

where $\bar{r}$ is the radius vector $\vec{OM}$, and $\bar{r}' = \vec{OM'}$. Indeed, (4.3.9) defines a point M' which satisfies the conditions of Definition 4.3.11.

Using formula (4.3.9), the reader can verify easily the following properties of an inversion: i) all the points of the circle (sphere) of center O and radius $\sqrt{k}$ are fixed points, if $k > 0$; ii) if $k < 0$, there are no fixed points, but the circle (sphere) of center O and radius $\sqrt{|k|}$ is fixed (this is the *circle (sphere)* of the inversion); iii) an inversion sends a straight line (plane) through the pole onto the same line (plane), and a line (plane) which does not contain the pole into a circle (sphere) through the pole. It is also possible to show that an inversion preserves the measure of angles. (Transformations which preserve angles are called *conformal transformations.*)

Exercises and Problems

4.3.1. Find the fixed points of the plane motion

$$x' = x\cos\varphi - y\sin\varphi + x_0,$$
$$y' = x\sin\varphi + y\cos\varphi + y_0.$$

(Orthogonal coordinates.)

4.3.2. Consider the triangle with vertices $A(0,0)$, $B(b,0)$, $C(b/2, b\sqrt{3}/2)$ ($b > 0$) (orthogonal coordinates). Show that there exists a rotation center and angle such that the rotation sends A to B, B to C and C to A. Write down the equations of this rotation. Is the same result true for any triangle?

4.3.3. In plane, the square $ABCD$ is rotated to the square $AB'C'D'$ by a rotation of angle φ around A. Prove that BB', CC', DD' are concurrent. Find the locus of $BB' \cap DD'$ if φ varies.

4.3.4. In plane, consider the transformation

$$x' = x\cos\theta + y\sin\theta + x_0,\ y' = x\sin\theta - y\cos\theta + y_0.$$

Show that it is an indirect orthogonal transformation, and find a decomposition of this transformation into a composition of a symmetry with respect to an axis and a translation in the direction of this axis.

4.3.5. In plane, let σ_1, σ_2 be symmetries with respect to the concurrent lines d_1, d_2. When is $\sigma_1 \circ \sigma_2 = \sigma_2 \circ \sigma_1$?

4.3.6. Find all the symmetry planes of a cube, and write down the equations of the symmetries with respect to these planes, and of their pairwise compositions, for an orthogonal frame with origin at a vertex of the cube and axes in the directions of the edges through this vertex. (A plane α is called a *symmetry plane* of a figure if the symmetry with respect to α preserves the figure as a whole.)

4.3.7. Let $(O, \bar{\imath}, \bar{\jmath}, \bar{k})$, $(O, \bar{\imath}', \bar{\jmath}', \bar{k}')$ be two positive, orthonormal frames, and let $\bar{\xi}$ be a unit vector on the intersection line of the planes $(O, \bar{\imath}, \bar{\jmath})$, $(O, \bar{\imath}', \bar{\jmath}')$ and such that the oriented angle from $\bar{\imath}$ to $\bar{\xi}$ is less than π, and, then, $\bar{\eta} = \bar{k} \times \bar{\xi}$. Consider the oriented angles $\varphi = \widehat{\bar{\imath}, \bar{\xi}}$, $\psi = \widehat{\bar{\xi}, \bar{\imath}'}$, $\theta = \widehat{\bar{k}, \bar{k}'}$, called it the Euler angles. Prove that the composition of the rotations: a) a rotation of angle φ around the axis $O\bar{k}$; b) a rotation of angle θ around the axis $O\bar{\xi}$; c) a rotation of angle ψ around the axis $O\bar{k}'$ sends $(\bar{\imath}, \bar{\jmath}, \bar{k})$ to $(\bar{\imath}', \bar{\jmath}', \bar{k}')$. Write down the resulting equations of this total rotation.

4.3.8. Prove, that if (x, y, z) are orthogonal coordinates, the formulas

$$x' = \frac{2}{3}x + \frac{2}{3}y + \frac{1}{3}z,$$

$$y' = \frac{11}{15}x - \frac{10}{15}y - \frac{2}{15}z,$$

$$z' = \frac{2}{15}x + \frac{5}{15}y - \frac{14}{15}z,$$

define a rotation around a point and find the axis of this rotation.

Chapter 5

Projective Geometry

5.1 Projective Incidence Properties

If you place yourself at the middle, between the parallel rails of a long straight railway track (when no train is arriving!), it seems that the rails meet somewhere, far away. In other words, our intuition finds an intersection point of two parallel lines *at infinity*. For the moment, this is just a way of speaking, since in the usual Euclidean (affine) geometry no notion of point at infinity is defined. In Euclidean space, what really happens is that two parallel lines have a common unmarked direction (see Section 1.1), and this is the explanation of our intuition of points at infinity.

Projective Geometry is the theory which translates the intuitive feeling described above into a logical, mathematical construction. This theory is very important for both pure mathematics and applications (e.g., perspective of drawing and painting, computer graphics). While results of projective geometry existed long ago, (Pappus - 4th century AD, Desargues - 17th century, etc.), the theory was developed extensively in the 19th century (Poncelet, Plücker, Grassmann, Klein and others), and this development was continued and is going on under various, more sophisticated aspects, and in its applications. The aim of this chapter is to give an introduction to projective geometry.

5.1.1 Definition. *i) An unmarked direction of the affine (Euclidean) space is called an improper point or an infinite point or a point at in-*

finity. ii) The locus of the improper points of all the straight lines of a given plane is called the improper straight line or the line at infinity of the plane. iii) The locus of the improper points of all the lines of affine (Euclidean) space is called the improper plane or the plane at infinity of space.

5.1.2 Definition. *A straight line with its added improper point is called an enlarged affine (Euclidean) straight line. A plane with its added improper line is called an enlarged affine (Euclidean) plane. Space with its added improper plane is called enlarged affine (Euclidean) space.*

Now, it is important to emphasize that Definitions 5.1.1, 5.1.2 also include a new definition of *belonging*. This notion has the classical meaning for *proper or finite* (as opposite to *improper or infinite*) points, lines and planes, and it is given by Definition 5.1.1 in the *improper* cases. Thus, if we agree to denote improper elements by an index ∞, a point A_∞ is an unmarked direction, and it is *represented* by a proper line a, and, if d is an arbitrary proper line and α is a proper plane, $A_\infty \in d$ iff $a \| d$, and $A_\infty \in \alpha$ iff $a \| \alpha$. Then, if d_∞ is an improper line, *represented* by a proper plane α (i.e., d_∞ is the improper line of α), $A_\infty \in d_\infty$ iff $a \| \alpha$. Finally, always $A_\infty \in \alpha_\infty$, where α_∞ is the (unique) improper plane of space. Similarly, d_∞ always is in α_∞, and d_∞ belongs to a proper plane β if the representing planes α of d_∞ are parallel to β. Notice that a proper line cannot have two distinct improper points, and a proper plane cannot have three noncollinear improper points. In geometry, the relation of *belonging* is also called *incidence.* That is, if either a point belongs to a line or a line passes through a point, we say that the point and the line are *incident*, and so on.

The enlarged, affine line, plane and space are said to be *models* of the *(real) projective line, projective plane and projective space*, respectively. What is meant by this expression is that projective space can be defined as the enlarged, affine space but, that we are ready to look at any other mathematical system of objects, which may be labeled as points, lines and planes, and which have the geometric properties encountered in the enlarged, affine space, as being projective space as well. This means that we are interested only in the geometric properties and not in the "concrete" definition of points, lines and planes.

Projective space is defined up to the equivalence of its geometric properties only, and these properties may be obtained on any of its models. For the moment, the only model which we have is the enlarged, affine space but, later on we will also study other models. We adopt this abstract point of view because there are many important models of the projective space. As a matter of fact, such abstract definitions are very common in mathematics, and we could have proceeded similarly with affine and Euclidean space. We actually did so when using coordinates, since the latter may be seen as a new model of space where a point is an ordered triple of real numbers (x, y, z), a plane is an equation $ax + by + cz + d = 0$, and a straight line is a pair of such equations. But, the intuitive geometric model of Euclidean space was preferred in the beginning, which is why we avoided to speak of models until now. Of course, what we just said about space holds for the projective line and plane as well.

Now, we have to be more precise in telling what kind of properties are considered in the geometry of projective space. Namely, *these will be the properties which are common to all the points, lines or planes, whether proper or improper in the enlarged space model.* In fact, in other models no such distinction between proper and improper will occur. These are called *projective.properties*, and the study of projective properties is *projective geometry.*

In particular, the study of *projective incidence properties* is an important part of projective geometry. The following propositions provide the *basic* projective incidence properties such that, in fact, any other projective incidence properties are consequences of these basic properties.

5.1.3 Proposition. *The following projective incidence properties are true:*
1. Any two distinct points belong to a unique straight line.
2. Any three noncollinear points belong to a unique plane.
3. If two distinct points of a straight line belong to a plane, any other point of the line also belongs to that plane.
4. If two planes have a common point, they must have at least another common point.
5. Every plane has at least one point. Every straight line has at least

three points. There exist three noncollinear points. There exist four noncoplanar points.
6. Two coplanar straight lines have a common point.

5.1.4 Proposition. *The following projective incidence properties are true:*
1′. Any two planes have a unique, common, straight line.
2′. Three planes which have no common line have a unique, common point.
3′. If M is a common point of two planes which pass through a straight line d, any plane through d contains M.
4′. If there exists one plane which passes through two points, there exists at least another plane through these two points.
5′. There exists at least one plane through any given point. There exist at least three planes through any given straight line. There exist three planes with no common straight line. There exist four planes with no common point.
6′. Any two straight lines with a common point are coplanar.

Proof. These properties may seem awkwardly trivial. This is because we chose them such that they are very basic, and play the role of *axioms* in an axiomatic construction of projective geometry. Here, these properties must be checked against the definitions of incidence given for the enlarged affine space. Even so, all these checks are very simple, and we leave them to the reader. The principle is to take into account all the possibilities for proper and improper elements, and see that the results hold in all the cases. The formulation of the results contains no distinction between proper and improper as required for truly projective properties.

For instance, for property 1, one has to look at two proper points A, B, and, then, they define a proper line only; at A, B_∞, and, then, the line which we need is the parallel through A to a line b which has B_∞ as the improper point; and at A_∞, B_∞ (i.e., the line is the improper line of a plane which is parallel to a, b, where the latter are lines with the improper points A_∞, B_∞). (See Fig. 5.1.1.)

We would like to draw the reader's attention to property 6 which, in fact, tells us that *the notion of parallel lines does not exist in projective geometry.* Indeed, two coplanar proper lines which have no common proper point are parallel from the affine point of view. But then, they have the same point at infinity, etc. Q.E.D.

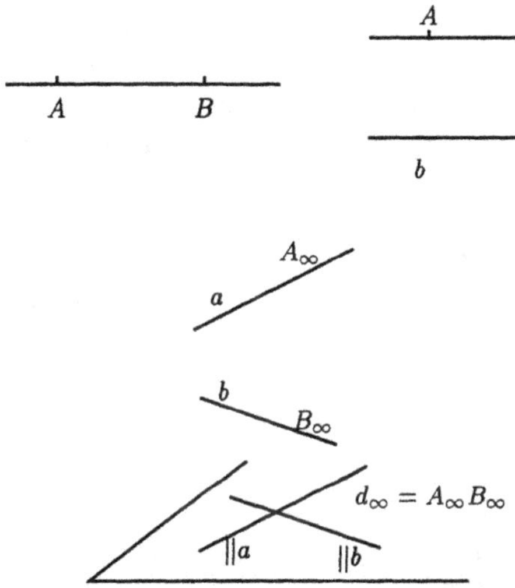

Fig. 5.1.1

Propositions 5.1.3 and 5.1.4 show a very important feature of projective geometry. Namely, the properties $1'$–$6'$ are obtained from 1–6 by a formal *translation* of the word *point* by the word *plane*, while a *straight line* remains a *straight line*, and an *incidence relation* remains an *incidence relation.* (Incidence may be expressed by various words, but only the significance of these words counts.) And, conversely, 1–6 are the translation of $1'$–$6'$ by the same rules. This translation is called *space duality.* On the other hand, one may see that the properties $1'$–$6'$ are logical consequences of 1–6, and conversely, and the projective incidence properties are exactly the logical consequences of either 1–6 or $1'$–$6'$. Hence, we get

5.1.5 Theorem. *(The Space Duality Principle) The dual of any pro-*

jective incidence theorem by space duality is again a theorem. (In other words, it is enough to either prove or disprove one of the two space-dual formulations, and, then, both of them will be simultaneously true or false, respectively.)

Furthermore, let us isolate those properties 1–6, which refer to plane geometry. These are 1, the second and third assertions of 5, and 6, and we may notice that 6 is essentially obtained from 1 by replacing the word *point* by *straight line*, and conversely, while keeping the incidence relations unchanged. This kind of translation is called *plane duality.* We see that the plane dual of 6 (plus the uniqueness of the intersection point) is 1. The plane duality translation of the second and the third part of 5 tell us that, in plane, there are always three lines through a given point, and, that there exist three nonconcurrent lines, and these assertions follow from 1 and 5. Since all the incidence properties of the projective plane geometry follow from the basic properties of 1, 5, 6, we get

5.1.6 Theorem. *(The Plane Duality Principle) The dual of any projective incidence theorem by plane duality is again a theorem of plane geometry. (In other words, it is enough to either prove or disprove one of the two plane-dual formulations, and, then, both of them will be simultaneously true or false, respectively.)*

Later on, we will give more justification for the projective duality principles, and, also, see that, in fact, they encompass the entire real projective geometry, not just incidence.

We do not make an in-depth study of projective incidence but, we will give one very important theorem

5.1.7 Theorem. *(Desargues' Theorem) In projective space, let $\triangle ABC$ and $\triangle A'B'C'$ be two triangles such that the lines AA', BB', CC' meet at a point O. Then, there exist intersection points $P = AB \cap A'B'$. $Q = BC \cap B'C'$, $R = CA \cap C'A'$, and these points are collinear. Conversely, if P, Q, R exist and are collinear, the lines AA', BB', CC' have a common point O.*

Proof. Two triangles as in the theorem are said to be *in perspective, with center O and axis d*, where d is the line through P, Q, R. That is, we have to prove that the existence of the center implies the existence of the axis and conversely.

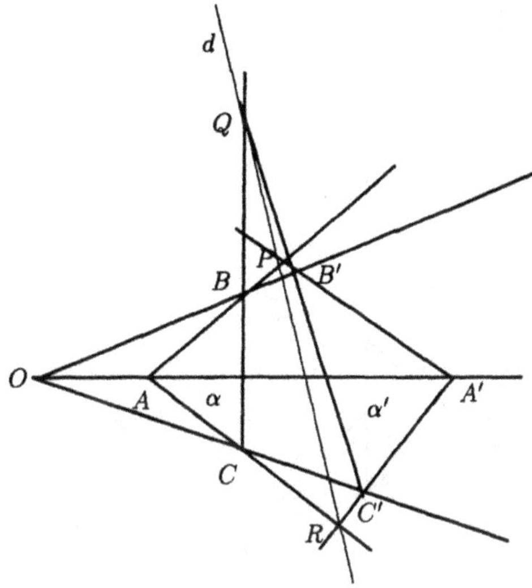

Fig. 5.1.2

First, we will assume that the two triangles are situated in different planes α and α', respectively (see Fig. 5.1.2). Then, AB, $A'B'$ belong to a common plane namely, OAB, and this implies the existence of P (see Proposition 5.1.3, 6). Similar arguments show the existence of Q and R. But, it is clear that P, Q, R are common points of the planes α, α', and, since two different planes meet along a straight line, P, Q, R will belong to this line, as required.

Now, let the two triangles belong to a single plane α; then the existence of P, Q, R is automatic. However, to see that they are collinear, we must get out of α. Namely, consider two more *centers* O_1, O_2, outside α, and such that O, O_1, O_2 are collinear (see Fig. 5.1.3). Then, if we look at the planes through these points and through A, B, C, we see that there exists a new triangle with the vertices $A'' = O_1A \cap O_2A'$, $B'' = O_1B \cap O_2B'$, $C'' = O_1C \cap O_2C'$. This triangle lies in a plane $\alpha' \neq \alpha$, and it is in perspective with the given triangles, with centers O_1, O_2, respectively. Hence, by the already proven case of the theorem,

we have the triples of collinear points

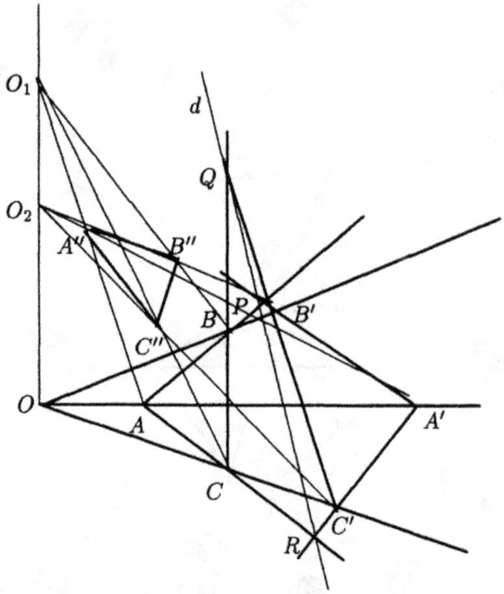

Fig. 5.1.3

$$P_1 = AB \cap A''B'', \quad Q_1 = BC \cap B''C'', \quad R_1 = CA \cap C''A'',$$

$$P_2 = A'B' \cap A''B'', \quad Q_2 = B'C' \cap B''C'', \quad R_2 = C'A' \cap C''A''.$$

Now, the lines AB, $A'B'$, $A''B''$ are three straight lines which do not belong to the same plane, and intersect pairwisely; this cannot happen unless all these lines have a common point, and the conclusion is that $P = P_1 = P_2$. In the same way, $Q = Q_1 = Q_2$, $R = R_1 = R_2$, and, therefore, the points P, Q, R belong to a common straight line, as required.

Conversely, the existence of the perspective axis implies the existence of the center is easy to prove by contradiction. If the axis exists, the lines AA', BB' belong to a plane, and have an intersection point O. OC must intersect $B'C'$ at some point C'', and $\triangle ABC, \triangle A'B'C''$ are in perspective from O (see Fig. 5.1.4). A contradiction is avoided iff the axis of these new triangles coincides with the axis of the given triangles, i.e., $C'' = C'$. Q.E.D.

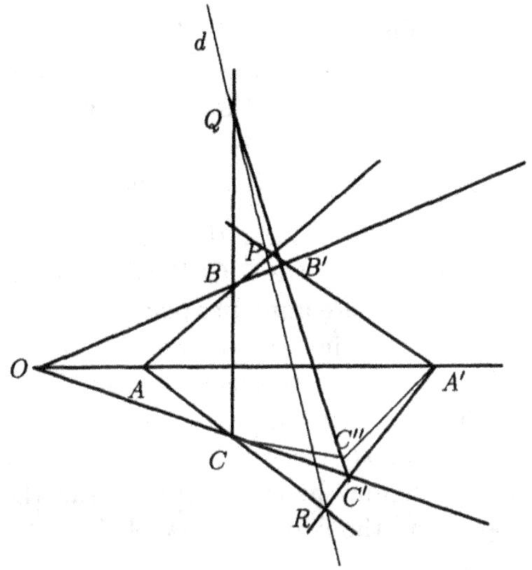

Fig. 5.1.4

5.1.8 Remarks. *1) It was not by chance that the proof of Desargues' theorem for two triangles in the same plane asked for a construction outside the plane. Indeed, there are models of the projective plane i.e., mathematical systems with points and lines which have the axiomatic plane projective incidence properties 1, 5, 6 of Proposition 5.1.3, and where the plane Desargues' theorem is not true (e.g., [14]). The explanation is that such systems cannot be embedded into larger systems which would be models of the projective space. A projective plane where Desargues' theorem does not hold is called a* nonarguesian plane. *2) The plane dual of the direct part of the plane Desargues' theorem is exactly its converse part. Therefore, this part actually doesn't need a special proof; it is true by the plane duality principle. This is not the case for space duality, however, and we suggest the reader to formulate by himself the space duals of the direct and the converse parts of Desargues' theorem.*

Desargues' theorem 5.1.7 is a projective theorem. If we use it in the affine enlarged space, and differentiate between proper and improper elements, we will find various affine theorems. For instance

5.1.9 Theorem. *(The Affine Desargues' Theorem) Let* $\triangle ABC$ *and* $\triangle A'B'C'$ *be two triangles of affine space, such that: i)* AA', BB', CC' *meet at a point* O*, ii)* $AB \| A'B'$, $AC \| A'C'$*. Then, we must also have* $BC \| B'C'$*. Conversely, if* $AB \| A'B'$, $AC \| A'C'$, $BC \| B'C'$*, the lines* AA', BB', CC' *either have a common point or are parallel.*

Proof. If we enlarge the affine space, the intersection points $P = AB \cap A'B'$, $Q = BC \cap B'C'$, $R = CA \cap C'A'$ exist, and are collinear. But, by hypothesis ii) P and R are improper points. Therefore, the line which joins them is an improper line, and, necessarily, $BC, B'C'$ have an improper intersection point Q; this means that $BC \| B'C'$, as required. The converse result follows similarly from the converse Desargues' theorem: the existence of the perspective axis implies the existence of the center but, now, the center may be either proper, and the respective lines meet, or improper, and the lines are parallel. Q.E.D.

Exercises and Problems

5.1.1. In enlarged affine space, prove that for any pair (a, b) of skew lines and for any point A which does not belong to a, b there exists one and only one straight line c through A which meets both a and b.

5.1.2. Prove that if $\triangle ABC$, $\triangle A'B'C'$, $\triangle A''B''C''$ are pairwise perspective with the same axis, the three perspective centers are collinear. Formulate the plane dual of this result.

5.2 Homogeneous Coordinates

In affine space, with respect to an affine frame, a point is represented by its affine coordinates (x, y, z), and an unmarked direction d is represented by a class of proportional vectors $\lambda\bar{v}(\lambda v_1, \lambda v_2, \lambda v_3)$, where $0 \neq \lambda \in \mathbf{R}$ and $\bar{v} \neq \bar{0}$. Since both the point and the direction are *projective points* (proper point and improper point of the enlarged space, respectively), we would like to unify their coordinate representation. This can be achieved as follows. Replace the triple of numbers (x, y, z) by the quadruple (x_1, x_2, x_3, x_0) (pay attention to the order!), which are related to (x, y, z) by the formulas

$$(5.2.1) \qquad x = \frac{x_1}{x_0}, \quad y = \frac{x_2}{x_0}, \quad z = \frac{x_3}{x_0}, \quad (x_0 \neq 0).$$

We warn the reader not to confuse the x_i $(i = 1, 2, 3, 0)$ of (5.2.1) with the notation $x = x_1$, $y = x_2$, $z = x_3$ used sometimes in previous chapters, and which will not be used anymore hereafter! Obviously, the numbers x_i of (5.2.1) are defined only up to multiplication with an arbitrary, nonzero scalar λ. We will say that these numbers are *homogeneous coordinates* of the proper point under discussion, with respect to the affine frame, while (x, y, z), were *nonhomogeneous coordinates.* Homogeneity just means the possibility of multiplying the coordinates by a common factor without changing the point.

On the other hand, if we have the unmarked direction d mentioned above, we will say that it defines the improper point of *homogeneous coordinates* $x_1 = \lambda v_1, x_2 = \lambda v_2, x_3 = \lambda v_3, x_0 = 0$. In this way, every point of the enlarged affine space, either proper or improper, has four homogeneous affine coordinates, which do not vanish simultaneously, and may be multiplied by an arbitrary nonzero scalar number. This unifies the analytical representation of the proper and improper points which, however, are differentiated by the fact that $x_0 \neq 0$ for the proper points, and $x_0 = 0$ for the improper points. Conversely, any ordered quadruple of homogeneous coordinates obviously belongs to a unique point of the enlarged space.

Similar considerations can be made for an enlarged affine plane. Then, we use only the first two formulas (5.2.1), and we get three homogeneous coordinates (x_1, x_2, x_0) with $x_0 \neq 0$ for proper points and $x_0 = 0$ for improper points. And, in the case of an enlarged affine line, we get two homogeneous coordinates (x_1, x_0). (Again, please pay attention to the order of the coordinates!)

Now, it is easy to establish analytical representations of various geometric figures in homogeneous coordinates starting from the usual affine representation. Thus, in plane, a proper straight line has an equation $a_1x + a_2y + a_0 = 0$, and, with (5.2.1), this equation becomes

$$(5.2.2) \qquad a_1x_1 + a_2x_2 + a_0x_0 = 0,$$

where (x_1, x_2, x_0) are homogeneous coordinates of an arbitrary proper point of the line. But, we may notice that (5.2.2) is also satisfied by the improper point of the same line since this improper point has the homogeneous coordinates $(-a_2, a_1, 0)$ (why?). Since the equation $x_0 = 0$ of the improper line of the plane is just a particular case of (5.2.2), we found that: *the general equation of an enlarged straight line of the enlarged affine (Euclidean) plane with respect to homogeneous affine coordinates is a homogeneous, linear equation of the form (5.2.2), where not all the coefficients a_i $(i = 1, 2, 0)$ vanish.*

Similarly, it follows that: *the general equation of an enlarged plane of enlarged affine (Euclidean) space is a homogeneous, linear equation of the form*

$$a_1x_1 + a_2x_2 + a_3x_3 + a_0x_0 = 0, \tag{5.2.3}$$

where not all the coefficients a_i $(i = 1, 2, 3, 0)$ vanish. (In order to check this equation for an improper point of the plane, we represent this improper point by a vector $\bar{v} = \vec{MN}$, where M, N belong to the plane; then, the homogeneous coordinates of the point will be $(v_1 = x_N - x_M,\ v_2 = y_N - y_M,\ v_3 = z_N - z_M, 0)$, and they satisfy (5.2.3).) Then, of course, a straight line may be seen as the intersection of two planes, *and it is represented by a pair of nonproportional equations* (5.2.3). In particular, the improper plane has the equation $x_0 = 0$, and an improper line, is the intersection of this plane with some proper plane.

And, to complete this picture, if we transform equations (3.3.1) and (3.3.2) of a conic (quadric) by (5.2.1), we get a general homogeneous, quadratic equation

$$\sum_{i,j} a_{ij}x_ix_j = 0, \tag{5.2.4}$$

where the indices i, j run from 0 to 2 in the case of a conic, and from 0 to 3 in the case of a quadric, and $a_{ij} = a_{ji}$. Thus, an equation of the form (5.2.4) represents an *enlarged conic (quadric)* i.e., one to which we add all the improper points which satisfy the equation.

We know how the nonhomogeneous coordinates change by a change of the affine frame. Namely, we have (1.3.10), which, now, we write with new coefficients, under the form

$$\tilde{x} = \alpha_{11}x + \alpha_{12}y + \alpha_{13}z + \alpha_{10}$$

$$\tilde{y} = \alpha_{21}x + \alpha_{22}y + \alpha_{23}z + \alpha_{20}$$
$$\tilde{z} = \alpha_{31}x + \alpha_{32}y + \alpha_{33}z + \alpha_{30}.$$

Then, in order to get the usual representation of the improper points with respect to both frames, we must consider that $x_0 = \lambda\tilde{x}_0$, where λ is any nonzero number, and, if we use (5.2.1) for the new coordinates too, we get

$$\lambda\tilde{x}_i = \sum_{j=0}^{3} \alpha_{ij}x_j, \tag{5.2.5}$$

where the coefficients are such that the equations may be solved with respect to x_i (linear algebra teaches us that this amounts to the fact that $det(\alpha_{ij}) \neq 0$), and

$$\alpha_{01} = \alpha_{02} = \alpha_{03} = 0. \tag{5.2.6}$$

A change of coordinates in plane has the same expression (5.2.5) with indices running up to 2, and on a line the indices are $0, 1$.

Now, we will make an important conceptual advancement. Namely, since we are not interested in differentiating between proper and improper points in projective geometry, it is natural to give the following definition

5.2.1 Definition. *Let (x_i) be affine homogeneous coordinates, and consider a general transformation of the form*

$$\lambda y_i = \sum_{j} \alpha_{ij}x_j, \tag{5.2.7}$$

where the coefficients are such that the system can be solved with respect to x_i (i.e., $det\,(\alpha_{ij}) \neq 0$), but we do not ask that (5.2.6) necessarily holds. Then, we get a one-to-one correspondence between points (proper and improper) and sequences (λy_i) $(i = 1, 2, 3, 0)$, where not all y_i vanish, and λ is an arbitrary nonzero scalar number. The (y_i) will be called a system of projective coordinates of space, plane or a line, respectively.

The projective coordinates are homogeneous coordinates (because of the arbitrary factor λ), and, at this moment, they are not related to a geometric figure, like the axes of an affine frame, but defined algebraically by means of a fixed transformation (5.2.7). Via this transformation, we may think of the projective coordinates as related to the affine frame of the coordinates x_i but, this is of not much use.

It is easy to understand the form of the equations of a straight line, plane, conic or quadric with respect to general projective coordinates. Namely, if (5.2.7) is solved for x_i, we get something of the form

$$\mu x_i = \sum_j \beta_{ij} y_j,$$

and the replacement of these solutions in equations (5.2.2), (5.2.3), (5.2.4) leads to similar equations with respect to y_i (with new coefficients, of course). Therefore, *the projective equations of straight lines, planes, conics, quadrics look exactly the same as their affine, homogeneous equations*, with the exception that $y_0 = 0$ doesn't necessarily represent improper points anymore.

Furthermore, if $(\tilde{y}_i)$ is another system of projective coordinates, obtained from the affine coordinates $(\tilde{x}_i)$, the corresponding transformation of projective coordinates is a composition of transformations of the form (5.2.7) and (5.2.5), and it gives a result of the same form

$$\rho \tilde{y}_i = \sum_{j=0}^{3} p_{ij} y_j, \tag{5.2.8}$$

with an arbitrary $\rho \neq 0$, and with coefficients such that $det(p_{ij}) \neq 0$ (i.e., the transformation is *invertible*; the equations can be solved for the y_i). (5.2.8) is a *general change of projective coordinates.* In matrix form, if we denote by P the matrix (p_{ij}) and if we write the coordinates in one-column matrices, (5.2.8) is

$$\rho \tilde{y} = P y. \tag{5.2.8'}$$

Finally, one more notational matter, since only the various quotients of the homogeneous coordinates are well defined, it is usual to denote such coordinates by $[y_1 : y_2 : y_3 : y_0]$, (in plane, $[y_1 : y_2 : y_0]$, on a line $[y_1 : y_0]$, and, again, pay attention to the order!).

Now, we can make one more step, and define

5.2.2 Definition. *Any mathematical system which consists of elements called "points", "lines" and "planes" such that the "points" can be put in a one-to-one correspondence with real "homogeneous coordinates"* $[y_1 : y_2 : y_3 : y_0]$, *the "planes" consist of sets of points which satisfy an equation (5.2.3), and the "lines" satisfy a nonproportional pair of equations (5.2.3) is called a model of the real three-dimensional projective space. The projective space itself, to be denoted by* $\mathbf{RP}^3$, *is defined as the abstract structure which underlies all its (equivalent!) models. The projective plane* $\mathbf{RP}^2$, *and the projective line* $\mathbf{RP}^1$ *are defined similarly. This definition can be generalized by saying that* $n+1$ *homogeneous coordinates* $[y_1 : y_2 : \ldots : y_n : y_0]$ *define the real, n-dimensional, projective space* $\mathbf{RP}^n$. *Similarly, if we use complex homogeneous coordinates, rather than real, we obtain the complex, projective spaces* $\mathbf{CP}^n$.

As a matter of fact, following Definition 5.2.2, we may always use as a "standard model" of the projective space the "numerical model"

(5.2.9)
$$\mathbf{RP}^n = \{[y_1 : y_2 : \ldots : y_n : y_0]\} = (\mathbf{R}^{(n+1)} \backslash \{0\})/(\textit{proportionality}),$$

where the last term is just another notation of the second term. (This notation says that $[y_1 : y_2 : \ldots : y_n : y_0]$ are the *equivalence classes* of correspondingly ordered sequences of $n+1$ real numbers which are not all zero, where the equivalence relation is the proportionality of two sequences. A similar notation is to be used for the complex space, with **C** instead of **R**.)

In our context, it was natural to start with the model of the enlarged affine space, and deduce the numerical model from the former. Now, we can give a few more models. If we look at the projective equation (5.2.2) of a straight line of the projective plane, we see that the line is determined by the coefficients (a_1, a_2, a_0), which are not all zero, and are defined up to proportionality (the line is not changed by a multiplication of its equation with a nonzero scalar). Hence, if we define as new points the lines of the projective plane, and as new lines the old points of the same plane, we obviously get a new model of the projective plane. *This model gives the full justification of the plane duality principle promised in Section* 5.1. Similarly (see equation (5.3.3), if we define as new points the old planes of the projective space, as new planes the old points of the projective space, and as new lines the old lines we obviously get a new model of projective space, and *this gives us the full justification of the space duality principle.*

On the other hand, if we define as new projective points the lines of an arbitrary pencil of straight lines of an affine plane, we obtain a new model of the projective line; the homogeneous coordinates of this model are the coefficients (λ, μ) of equation (2.3.18) of the pencil. The same thing can be done with a pencil of planes (2.3.19). As a matter of fact, pencils of concurrent lines can be defined in any projective plane, and pencils of planes through a line can be defined in projective space also. As in the affine case, such pencils are defined by a pair of elements of the pencil, and are represented by equations of the type (2.3.18), (2.3.19) in homogeneous, projective coordinates. Every such pencil is a model of the projective line again.

Similarly, if we take a bundle of planes of affine space or of projective space, defined by equation (2.3.25), and take its planes as new points, and its pencils of planes as new lines, we get a model of the projective plane, and we ask the reader to justify that a similar model of the projective plane is defined by a bundle of straight lines (i.e., the lines of space through a fixed point), etc.

We end this section by indicating the interesting fact that, since we now have a general definition of projective space, we can proceed in the converse direction, and deduce affine space from projective space. Consider the projective space $\mathbf{RP}^3$ by, say, its numerical model, and chose a certain plane α of this space (i.e., a certain equation (5.3.3)). Then, if we agree to see the points and the lines of α as improper points and lines, the space $\mathbf{RP}^3$ gets the structure of an enlarged affine space, and $\mathbf{RP}^3 \backslash \alpha$ gets the structure of an affine space. Two lines of this affine space will be parallel if they belong to a plane, and their projective intersection point belongs to α.

In order to get a better understanding of this fact, we make a projective change of coordinates (5.2.8) such that the new equation of the plane α will be $\tilde{y}_0 = 0$; this is always possible. Then, the points of $\mathbf{RP}^3 \backslash \alpha$ get *nonhomogeneous projective coordinates* $\tilde{x} = \tilde{y}_1/\tilde{y}_0$, $\tilde{y} = \tilde{y}_2/\tilde{y}_0$, $\tilde{z} = \tilde{y}_3/\tilde{y}_0$, which behave just like usual affine coordinates. This shows that $\mathbf{RP}^3 \backslash \alpha$ *is a model of affine space.* By enlarging this affine space, we regain the projective space $\mathbf{RP}^3$. In a similar way, generally, if we take out a *hyperplane* (i.e., a subset defined by one homogeneous, linear equation) from the projective space $\mathbf{RP}^n$, we obtain an *n-dimensional affine space.*

5.2.3 Remark. *As above, in projective geometry it may be useful to use nonhomogeneous, projective coordinates in other situations as well. Such coordinates can always be obtained on various portions of space, but not on the whole space. More exactly, we have* $\mathbf{RP}^n = \cup_{i=0}^{n} U_i$, *where* U_i *is the subset of* $\mathbf{RP}^n$ *where* $y_i \neq 0$, *and the quotients* y_j/y_i $(j \neq i)$ *are nonhomogeneous coordinates on the subset* U_i.

Exercises and Problems

5.2.1. Let $A_1A_2A_3A_4A_5A_6$ be a plane regular hexagon, and consider an affine frame with origin A_1 and vector basis $\bar{e}_1 = \overrightarrow{A_1A_2}$, $\bar{e}_2 = \overrightarrow{A_1A_3}$.
i) Compute the homogeneous coordinates of all the vertices of the hexagon and of the improper points of all its sides. ii) Write down the homogeneous equations of the sides of the hexagon.

5.2.2. A cube of edge length 1 is placed upward on the plane Oxy of an orthogonal frame of space so that its vertical edges are in Oxz, Oyz, respectively. Compute the homogeneous coordinates of the improper points of its edges, and show that these points are not collinear.

5.2.3. If $Oxyz$ is an affine frame of space, define all the systems of projective coordinates (y_i) $(i = 1, 2, 3, 0)$ such that, with respect to these coordinates the plane xOy has the equation $y_0 = 0$.

5.3 Cross Ratios and Projective Frames

We would like to know whether, despite the abstract definition of projective coordinates in the previous section, it wouldn't be possible to connect the projective coordinates with a specific kind of *geometric frame*. This will be achieved by means of the important notion of a *cross ratio*.

Let d be a projective line with a projective coordinate system (x_1, x_0) fixed on it. Let M_a $(a = 1, 2, 3, 4)$ be four points of d. Then, the *cross ratio or anharmonic ratio* of these points is the number defined by

$$(5.3.1)\qquad (M_1, M_2; M_3, M_4) = \frac{\begin{vmatrix} x_1^1 & x_0^1 \\ x_1^3 & x_0^3 \end{vmatrix}}{\begin{vmatrix} x_1^2 & x_0^2 \\ x_1^3 & x_0^3 \end{vmatrix}} : \frac{\begin{vmatrix} x_1^1 & x_0^1 \\ x_1^4 & x_0^4 \end{vmatrix}}{\begin{vmatrix} x_1^2 & x_0^2 \\ x_1^4 & x_0^4 \end{vmatrix}},$$

where the lower indices indicate the place of the coordinates and the upper indices indicate what point M_a is used.

In (5.3.1) we accept the possibility of two points being equal (i.e., only three of the four points are distinct), and, accordingly, the possibility of the cross ratio being ∞. The notion of a cross ratio is important because of

5.3.1 Proposition. *The cross ratio of four points of a projective line is invariant by projective changes of coordinates.*

Proof. A projective coordinate change has the form (5.2.8′) where P is a 2 by 2 matrix, and a simple computation shows that, for instance,

$$\rho^2 \begin{vmatrix} \tilde{x}_1^1 & \tilde{x}_0^1 \\ \tilde{x}_1^3 & \tilde{x}_0^3 \end{vmatrix} = det(P) \begin{vmatrix} x_1^1 & x_0^1 \\ x_1^3 & x_0^3 \end{vmatrix}.$$

If this formula is used for all the determinants of (5.3.1), we see that ρ and $det(P)$ cancel, and the expression of the cross ratio is the same for the two systems of projective coordinates. Q.E.D.

We emphasize that definition formula (5.3.1) must be understood in its full generality i.e., as a definition which may be used in any model of the projective line. In particular, we may speak of the cross ratio of four points on an arbitrary straight line of a projective plane or space. In order to see how to compute the cross ratio in this case we first prove the following result which is also important in its own right:

5.3.2 Proposition. *In projective space (plane, line), and with respect to a system of projective coordinates, if $M_a(x_i^a)$ $(a = 0, 1;\ i = 1, 2, 3, 0)$ are two points, then any point $M(x_i)$ of the line M_0M_1 can be expressed as*

$$(5.3.2)\qquad x_i = \lambda_0 x_i^0 + \lambda_1 x_i^1,$$

where λ_0, λ_1 are parameters defined up to multiplication with an arbitrary nonzero factor, and which do not vanish simultaneously.

Proof. Since we may go over from projective space to affine space by an arbitrary choice of the improper plane, we will do this in such a way that M_0 will be a proper point, and M_1 an improper point. Then, $M_0[x_0 : y_0 : z_0 : 1]$, where (x, y, z) are nonhomogeneous, affine coordinates, and $M_1[v_1 : v_2 : v_3 : 0]$, where, in fact, v_i are the affine direction parameters of the line M_0M_1. An arbitrary proper point of the line will be $M[x : y : z : 1]$, and it is well known from affine geometry that

$$x = x_0 + tv_1, \quad y = y_0 + tv_2, \quad z = z_0 + tv_3, \tag{5.3.3}$$

where t is a parameter. If the indicated homogeneous coordinates of M are multiplied by a nonzero factor, the equations (5.3.3) are exactly (5.3.2), with $\lambda_0 \neq 0$. And, obviously, the improper point M_1 is given by (5.3.2) again, but with $\lambda_0 = 0$. Hence, (5.3.2) is proven for projective coordinates associated with the mentioned choice of the improper plane. But, a projective change (5.2.7) of coordinates leaves the form of (5.3.2) unchanged. (Reader who is familiar with the general theory of homogeneous, linear equations may justify (5.3.1) from that theory, directly.) Q.E.D.

Notice that this result shows again that a straight line of projective space is a projective line; λ_0 and λ_1 are *internal projective coordinates* on this line. If we look at the case of a line from the beginning i.e., we only have the coordinates $[x_1 : x_0]$, we get $\lambda_0 = x_0$, $\lambda_1 = x_1$, if we start with the points $M_0[1 : 0]$, $M_1[0 : 1]$.

The equations (5.3.2) are called the *parametric equations* of the line through M_0, M_1.

Now, if we come back to the cross ratio, since the parameters λ of the parametric equation of a line are projective coordinates, the cross ratio of the points A_a $(a = 1, 2, 3, 4)$ of the line (5.3.2) is given by

$$(A_1, A_2; A_3, A_4) = \frac{\begin{vmatrix} \lambda_1^1 & \lambda_0^1 \\ \lambda_1^3 & \lambda_0^3 \end{vmatrix}}{\begin{vmatrix} \lambda_1^2 & \lambda_0^2 \\ \lambda_1^3 & \lambda_0^3 \end{vmatrix}} : \frac{\begin{vmatrix} \lambda_1^1 & \lambda_0^1 \\ \lambda_1^4 & \lambda_0^4 \end{vmatrix}}{\begin{vmatrix} \lambda_1^2 & \lambda_0^2 \\ \lambda_1^4 & \lambda_0^4 \end{vmatrix}}, \tag{5.3.4}$$

where the values of λ correspond to the representation (5.3.2) of the given points A_a. The cross ratio (5.3.4) is independent of the choice of the basic points M_0, M_1 since a change of these points results in a projective change of the parameters λ.

Furthermore, if we look at the other models of the projective line described towards the end of Section 5.2, we see that we may similarly define the cross ratio of four straight lines of a pencil i.e., four lines of a projective plane which have a common point, and of the cross ratio of four planes of a pencil of planes i.e., four planes of projective space which have a common straight line. In both cases, the expression of the cross ratio is (5.3.4) again, where, now, the parameters are those of the equations (2.3.18), (2.3.19) of the pencils. These cross ratios have the following important property

5.3.3 Proposition. *Let d_u $(u = 1,2,3,4)$ be four lines of a pencil situated in a projective plane. Let A_u be the intersection points of these lines with another straight line d of the plane. Then $(A_1, A_2; A_3, A_4) = (d_1, d_2; d_3, d_4)$. The same property holds for the intersection points of a line with four planes of a pencil in space.*

Proof. (See Fig. 5.3.1.) Let (x_1, x_2, x_0) be projective coordinates in the plane of the lines d_u, and let (a_i^u) $(i = 1,2,0;\ u = 1,2,3,4)$ be the coordinates of the points A_u. If A_1, A_2 are chosen as the basic points of the line d, we get $a_i^3 = \lambda a_i^1 + \mu a_i^2$, and $a_i^4 = \lambda' a_i^1 + \mu' a_i^2$, and formula (5.3.4) yields $(A_1, A_2; A_3, A_4) = \lambda'\mu/\mu'\lambda$. On the other hand, assume that the equations of d_1, d_2 are

$$d_1 := \alpha_1 x_1 + \alpha_2 x_2 + \alpha_0 x_0 = 0, \quad d_2 := \beta_1 x_1 + \beta_2 x_2 + \beta_0 x_0 = 0,$$

and that the other two lines of the pencil are

$$d_3 = \sigma d_1 + \tau d_2 = 0, \quad d_4 = \sigma' d_1 + \tau' d_2 = 0.$$

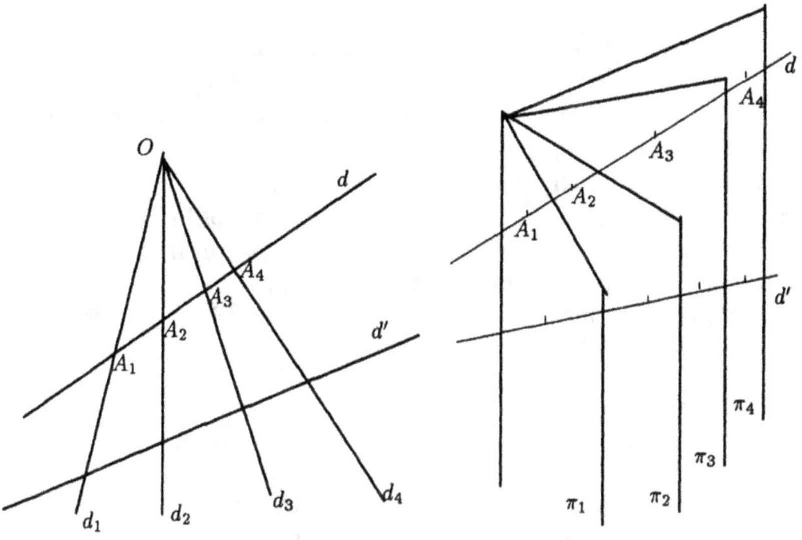

Fig. 5.3.1

Then, the same formula (5.3.4) yields $(d_1, d_2; d_3, d_4) = \sigma'\tau/\sigma\tau'$.

But, we have $A_u \in d_u$, $\forall u = 1,2,3,4$, i.e., $d_u(A_u) = 0$, where the notation means that we evaluate the left-hand side of the equation of the line d_u for the coordinates of the point A_u. Accordingly, we obtain

$$\sigma d_1(\lambda A_1 + \mu A_2) + \tau d_2(\lambda A_1 + \mu A_2) = \sigma\mu d_1(A_2) + \tau\lambda d_2(A_1) = 0,$$

$$\sigma' d_1(\lambda' A_1 + \mu' A_2) + \tau' d_2(\lambda' A_1 + \mu' A_2) = \sigma'\mu' d_1(A_2) + \tau'\lambda' d_2(A_1) = 0,$$

whence

$$\frac{\sigma}{\lambda d_2(A_1)} = \frac{\tau}{-\mu d_1(A_2)}, \quad \frac{\sigma'}{\lambda' d_2(A_1)} = \frac{\tau'}{-\mu' d_1(A_2)}.$$

From the last two equalities, and the values of the cross ratios computed earlier, the assertion of the proposition follows immediately. The computations which prove the proposition for four planes through a straight line are exactly the same. Q.E.D.

5.3.4 Corollary. *If four collinear points are projected from either another point or another line onto a straight line, the value of the cross ratio is preserved.*

Indeed, the cross ratio remains equal to that of the projecting lines or planes (see Fig. 5.3.1). The properties of Proposition 5.3.3 and Corollary 5.3.4 are expressed by saying that *the cross ratio is invariant by intersections and projections.*

In order to understand where this "complicated" notion of cross ratio comes from, we will calculate the cross ratio of four proper points of an enlarged Euclidean line d. If these points are M_a $(a = 1,2,3,4)$ as in (5.3.1), and if the homogeneous coordinates of (5.3.1) are defined from an orthonormal frame of the proper line d by $x_a = x_1^a/x_0^a$, the cross ratio (5.3.1) becomes

$$(5.3.5) \qquad (M_1, M_2; M_3, M_4) = \frac{x_3 - x_1}{x_3 - x_2} : \frac{x_4 - x_1}{x_4 - x_2}$$

$$= \frac{a(\overrightarrow{M_1M_3})}{a(\overrightarrow{M_2M_3})} : \frac{a(\overrightarrow{M_1M_4})}{a(\overrightarrow{M_2M_4})} = \frac{(M_1, M_2; M_3)}{(M_1, M_2; M_4)}.$$

Formula (5.3.5) relates the cross ratio with distances and simple ratios, and explains the name of a cross ratio.

On the other hand, there are many nice geometric properties of four points with cross ratio -1, and, in this case, the cross ratio and the four points are called *harmonic*. More exactly, one says that M_3, M_4 are *harmonically conjugated with respect to* M_1, M_2 and conversely. This explains the general name of *anharmonic ratio*. Notice also that, if we take $x_4 \to \infty$ in (5.3.5), we obtain

$$(M_1, M_2; M_3, M_\infty) = \frac{x_3 - x_1}{x_3 - x_2}$$

i.e., the cross ratio reduces to a simple ratio.

Now, we will use formula (5.3.5) in order to prove the general properties of a cross ratio

5.3.5 Proposition. *i) Given three distinct points of a projective line* M_1, M_2, M_3, *there exists one and only one point* M_4 *such that* $(M_1, M_2; M_3, M_4)$ *has an arbitrary given value* k. *ii) The following equalities hold for any four points of any model of the projective line:*

$$(5.3.6) \quad (M_1, M_2; M_3, M_4) = (M_2, M_1; M_4, M_3) = (M_3, M_4; M_1, M_2) =$$

$$= 1/(M_2, M_1; M_3, M_4) = 1 - (M_1, M_3; M_2, M_4).$$

iii) For any five points of a projective line one has

$$(M_1, M_2; M_3, M_4)(M_1, M_2; M_4, M_5)(M_1, M_2; M_5, M_3) = 1 \tag{5.3.7}$$

(the Möbius identity).

Proof. i) Since the elimination of an arbitrary point of a projective line leaves us with an affine line, we may assume that the given three points are proper points, and use (5.3.5), including the case $x_4 \to \infty$. And, (5.3.5) is uniquely solvable with respect to x_4, the possibility of the solution $x_4 = \infty$ included. ii) By the same argument, we may assume now that all four points are proper. Then, (5.3.6) follows by straightforward computations. iii) If the cross ratios of (5.3.7) are expressed by coordinates, everything cancels. Q.E.D.

5.3.6 Remark. *(5.3.6) means that the cross ratio remains unchanged if either the pairs (M_1, M_2), (M_3, M_4) are interchanged or the order in both these pairs is changed. And, if $(M_1, M_2; M_3, M_4) = k$, then*

$$(M_2, M_1; M_3, M_4) = (M_1, M_2; M_4, M_3) = 1/k, \tag{5.3.8}$$

$$(M_1, M_3; M_2, M_4) = (M_4, M_2; M_3, M_1) = 1 - k. \tag{5.3.9}$$

Accordingly, while it seems that one could define 24 cross ratios of four points, taken in all possible orders, only six of them may be different. Indeed, by operations indicated in Remark 5.3.6, it is always possible to bring M_1 on the first place without changing the value of the cross ratio. And, the six values which may be obtained are

$$k,\ \frac{1}{k},\ 1-k,\ \frac{1}{1-k},\ \frac{k}{k-1},\ \frac{k-1}{k}. \tag{5.3.10}$$

The number of distinct values is even smaller if the numbers (5.3.10) are not distinct. This happens if $k = \pm 1$, $k = 0$, $k = \infty$, $k = 2$, $k = 1/2$ and $k = (1 \pm i\sqrt{3})/2$. In the last case, the ratio is called *equianharmonic*; if $k = -1$, the ratio is *harmonic*; $k = 2, 1/2$ are values of harmonic quadruples put in a different order; $k = 0, 1, \infty$ are obtained if only three of the four points are distinct.

As we already said, there are a lot of nice geometric properties of cross ratios. We do not intend to study them here but, just to get their flavour, we give one such result. Four points of a projective plane are the *vertices* of a *complete quadrangle* ; the latter has six *sides* i.e., the lines which join pairs of vertices, and three *diagonal points* i.e., the crossing points of opposite sides (see Fig. 5.3.2, where A, B, C, D are the vertices, AB, AC, AD, BC, BD, CD are the sides, and E, F, G are the diagonal points). Then, we have

5.3.7 Theorem. *(Desargues) On every side of a complete quadrangle, the two vertices, the diagonal point, and the intersection with the line which joins the other two diagonal points form a harmonic ratio.*

Proof. With Fig. 5.3.2, we want to prove that $(B, C; E, X) = -1$. By projecting these points from G onto AB, and then from F back to BC, and since projections and intersections preserve the cross ratio, we get

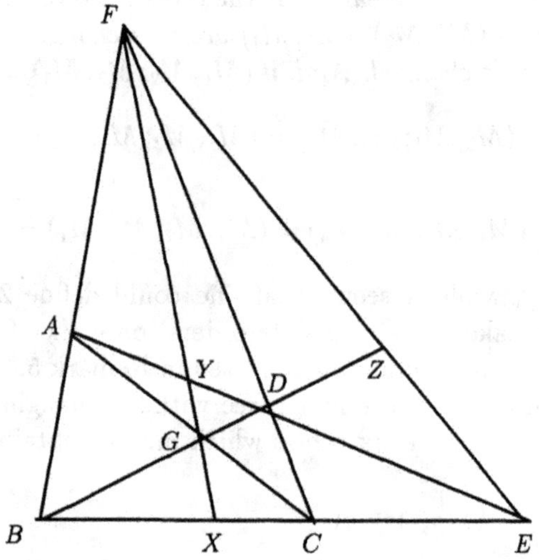

Fig. 5.3.2

$$(B, C; E, X) = (D, A; E, Y) = (C, B; E, X).$$

Since the first and last terms of this equality are inverse of each other, and since the four points are distinct, we must have the required value $(B, C; E, X) = -1$. Q.E.D.

We advise the reader to establish the plane dual of Theorem 5.3.7. The plane dual figure of a complete quadrangle is a *complete quadrilateral* i.e., the figure generated by four straight lines, the *sides*; it has six *vertices*, intersection points of pairs of sides, and three *diagonals*.

Now, we will address the problem of the geometric interpretation of the projective coordinates.

Let d be a projective line, and (y_1, y_0) an abstractly defined system of projective coordinates of d. Then, there are points of d which have the following y-coordinates $A_0[0 : 1]$, $A_1[1 : 0]$, and $U[1 : 1]$, and the triple of pairwise distinct points (A_0, A_1, U) is called the *projective frame* associated with the projective coordinates y_i $(i = 1, 0)$ (see Fig. 5.3.3). A_0, A_1 are called the *fundamental points* of the frame, and U is the *unit point*. If $M[y_1 : y_0]$ is an arbitrary point of d, formula (5.3.1) yields

(5.3.11) $$(A_1, A_0; U, M) = \frac{y_1}{y_0}.$$

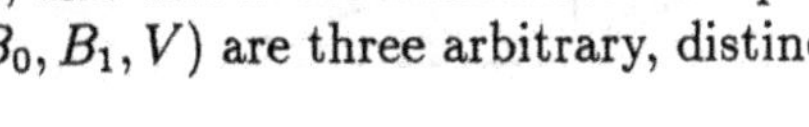

Fig. 5.3.3

Formula (5.3.11) shows that the nonhomogeneous projective coordinate on d is a cross ratio, and this is the announced interpretation.

Conversely, if (B_0, B_1, V) are three arbitrary, distinct points of d, if we define

(5.3.12) $$\frac{z_1}{z_0} = (B_1, B_0; V, M),$$

and if we compute this cross ratio using the y-coordinates of (B_0, B_1, V) and formula (5.3.1), we get relations of the form

(5.3.13) $$\rho z_i = \sum_{j=0,1} a_{ij} y_j,$$

where $\rho \neq 0$ is an arbitrary factor which appears because only z_1/z_0 is defined by (5.3.12). Moreover, (5.3.13) are invertible i.e., solvable with respect to y_1, y_0 because of Proposition 5.3.5 i). Hence, $[z_1 : z_0]$ is a new system of projective coordinates of d with associated fundamental points B_0, B_1, and unit point V.

Therefore, *the systems of projective coordinates of d are in a one-to-one correspondence with the projective frames, and the nonhomogeneous projective coordinates are cross ratios.*

Furthermore, let $[y_1 : y_2 : y_0]$ be projective coordinates of a projective plane. We will define the associated *fundamental points* by $A_1[1:0:0]$, $A_2[0:1:0]$, $A_0[0:0:1]$, and the *unit point* by $U[1:1:1]$, and we say that (A_1, A_2, A_0, U) is the projective frame associated with the coordinates (y_i). It is important to notice that no three of the four points of the frame are collinear (check!).

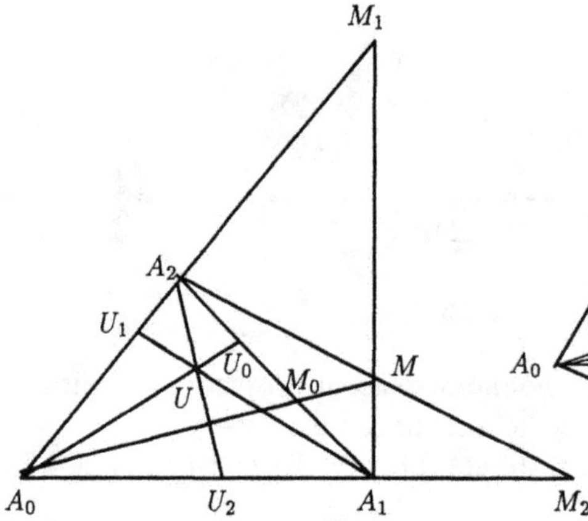

Fig. 5.3.4

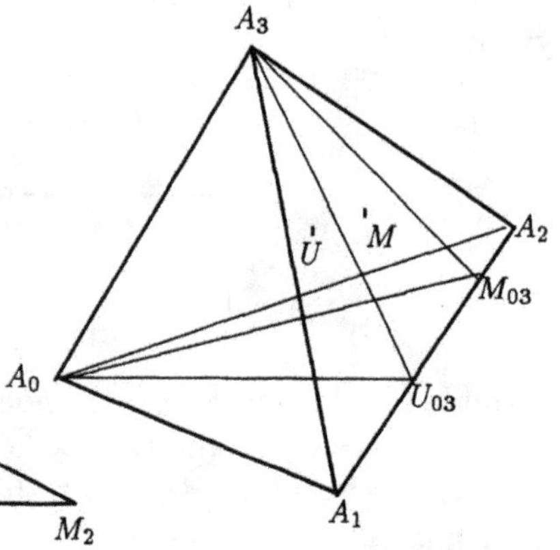

Fig. 5.3.5

Let $M[y_1 : y_2 : y_0]$ be any point of the plane. Using the invariance of the cross ratio by projections, and projecting from A_2 onto A_1A_0, from A_1 onto A_2A_0 and from A_0 onto A_1A_2 (Fig. 5.3.4), we get the interpretation of the nonhomogeneous coordinates of M as cross ratios:

(5.3.14) $$\frac{y_1}{y_0} = (A_1, A_0; U_2, M_2), \quad \frac{y_2}{y_0} = (A_2, A_0; U_1, M_1),$$

$$\frac{y_2}{y_1} = (A_2, A_1; U_0, M_0).$$

Conversely, the choice of a *projective frame* consisting of three fundamental points B_0, B_1, B_2 and a unit point V, such that no three of these four points are collinear, allows us to define new projective coordinates $[z_1 : z_2 : z_0]$ by the formulas (5.3.14) with A_i replaced by B_i and U replaced by V. This follows by the same arguments as in the case of the projective line. Moreover, the conclusion of these arguments is that *there exists a one-to-one correspondence between the systems of projective coordinates of the plane and the projective frames of the plane.*

Finally, in the same way, a *projective frame* in space is defined by a *fundamental tetrahedron* A_0, A_1, A_2, A_3 and a unit point U which does not belong to any of the planes of the faces of the tetrahedron (i.e., no four of these five points belong to the same plane), and the projective coordinates associated with this frame are defined by

$$\frac{y_2}{y_1} = (A_2, A_1; U_{03}, M_{03}), \qquad \text{etc.,} \tag{5.3.15}$$

where the notation is that of Fig. 5.3.5 i.e., U_{03}, M_{03} are the projections of U, M from the line A_0A_3 onto A_1A_2, etc. (The reader is advised to fill in the missing details.)

From the definition of the projective frame of a projective coordinate system it follows that, if we ask

$$U = \sum_i A_i, \tag{5.3.16}$$

where the relation is to be understood as holding for the respective coordinates of the implicated points, the relation between a point M and the points of frame may be written formally as

$$M = \sum_i y_i A_i, \tag{5.3.17}$$

where, again, (5.3.17) is to be understood as a relation between the respective coordinates. This remark allows us to calculate the projective coordinates z_i of a point M with respect to any projective frame

(B_i, V), where this new frame is defined by the homogeneous coordinates of its points with respect to a previously defined coordinate system y_i. Namely, we first fix the arbitrary factors allowed by the coordinates of B_i by solving the system

$$\sum \lambda_i B_i = V \tag{5.3.18}$$

with respect to the λ_i; the latter are determined up to an arbitrary, common factor ρ which is the factor admitted by the coordinates of V. Then, we change the coordinates of the fundamental points by taking $\tilde{B}_i = \lambda_i B_i$. Finally, we find the coordinates of M, up to the factor ρ, by solving z_i from the system

$$M = \sum_i z_i \tilde{B}_i. \tag{5.3.19}$$

This procedure emphasizes the role of the unit point of the frame. Without the unit point (5.3.19) becomes a mess because of too many arbitrary factors of the coordinates of the points there. (See Example 5.3.9 after the next Remark.)

5.3.8 Remark. *With linear algebra in mind, from formulas like (5.3.17), (5.3.19) we shall notice that it is possible to discuss linear combinations of points of a projective space, similar to the linear combinations of vectors of a linear space. This is a consequence of the representation* $\mathbf{RP}^n = (\mathbf{R}^{n+1}\backslash\{0\})/(proportionality)$, *since* $\mathbf{R}^{n+1}$ *is a linear space. Linear dependence and independence of points may be discussed in the usual way using the corresponding vectors of* $\mathbf{R}^{n+1}$. *For instance, Proposition 5.3.2 shows that three points are collinear iff they are linearly dependent points. But, the use of the coefficients of a linear combination is more complicated because of the arbitrary factors admitted by the projective points.*

5.3.9 Example. Let $[y_1 : y_2 : y_0]$ be plane projective coordinates. Take a new projective frame defined by

$$B_1[0:1:2],\ B_2[2:1:0],\ B_0[1:1:-1],\ V[0:0:1].$$

If the projective coordinates with respect to this frame are $[z_1 : z_2 : z_0]$, find the coordinate transformation between y_i, and z_i.

Solution. What we have to do is to compute exactly the z-coordinates of an arbitrary point M. Thus, we must first solve the system (5.3.18) which in our case is

$$\begin{aligned}\lambda_0.1+\lambda_1.0+\lambda_2.2 &= 0,\\ \lambda_0.1+\lambda_1.1+\lambda_2.1 &= 0,\\ \lambda_0.(-1)+\lambda_1.2+\lambda_2.0 &= 1.\end{aligned}$$

The solutions are: $\lambda_0=-1/2$, $\lambda_1=1/4$, $\lambda_2=1/4$.

Then, the required transformation is given by (5.3.19) which, now, is

$$\begin{aligned}\rho y_1 &= 0z_1+\frac{1}{2}z_2-\frac{1}{2}z_0,\\ \rho y_2 &= \frac{1}{4}z_1+\frac{1}{4}z_2-\frac{1}{2}z_0,\\ \rho y_0 &= \frac{1}{2}z_1+0z_2+\frac{1}{2}z_0,\end{aligned}$$

and it may be solved with respect to z_i $(i=1,2,0)$.

Exercises and Problems

5.3.1. The line

$$\frac{x-2}{1}=\frac{y-3}{2}=\frac{z-5}{6}$$

cuts the coordinate planes xOy, yOz, zOx at A, B, C, respectively. Compute the cross ratio $(A,B;C,P)$, where $P(2,3,5)$.

5.3.2. i) Let h_1,h_2,h_3,h_4 be four half lines which define the straight lines d_1,d_2,d_3,d_4 of a Euclidean plane. Then:

$$(d_1,d_2;d_3,d_4)=\frac{\sin\widehat{h_1h_3}}{\sin\widehat{h_2h_3}}:\frac{\sin\widehat{h_1h_4}}{\sin\widehat{h_2h_4}},$$

where the angles are oriented angles. ii) Let $A_1A_2A_3A_4A_5$ be a plane regular pentagon. Compute the cross ratio of four lines $(A_1A_2,A_1A_3;A_1A_4,A_1A_5)$.

5.3.3. The vertices of a pentagon of the affine plane are $A(0,1)$, $B(1,0)$, $C(1,1)$, $D(1/2,2)$, $E(1/4,2)$. Compute the cross ratio $(AB,AC;AD,AE)$.

5.3.4. Let A, B, C, D be four proper points of a Euclidean line. Show that $(A, B; C, D)$ is a harmonic ratio iff $MN^2 = MA^2 + NC^2$, where M is the midpoint of AB, and N is the midpoint of CD.

5.3.5. (*Pappus' Theorem*) Let d, d' be two straight lines of a projective plane, and the arbitrary points $A, B, C \in d$, $A', B', C' \in d'$. Prove that the intersection points $P = AB' \cap A'B$, $Q = AC' \cap A'C$, $R = BC' \cap B'C$ are collinear.

5.3.6. In plane, take two straight lines d_1, d_2 which meet at O, and a point P which belongs to none of them. A variable line d through P cuts d_1, d_2 in M, N, respectively. Find the locus of $Q \in d$ such that $(M, N; P, Q)$ is harmonic.

5.3.7. In the enlarged affine plane, let ΔABC be a triangle with proper vertices, and M its centroid. Find the equation of the improper line of the plane with respect to the projective frame with the fundamental points A, B, C, and the unit point M.

5.3.8. In a projective plane, take a quadrilateral $ABCD$, and let d be an arbitrary line through the intersection of the diagonals $U = AC \cap BD$. Denote: $X = AB \cap d$, $Y = CD \cap d$, $Z = BC \cap d$, $S = DA \cap d$. Prove that $\forall E \in d$ one has

$$(U, S; E, X) + (U, S; E, Y) = (U, S; E, Z).$$

(Hint: Use a plane projective frame with the fundamental points A, B, C and the unit point D.)

5.3.9. Let $ABCD$ be a tetrahedron, and d a straight line which intersects the faces BCD, CDA, DAB, ABC in A', B', C', D', respectively. Prove that

$$(A', B'; C', D') = ((d, A), (d, B); (d, C), (d, D)),$$

where (d, A) is the plane which contains d and A, and so on.

5.4 Conics, Quadrics, Projective Transformations

Some basic formulas of linear, analytical, projective geometry were established in the previous sections (formulas (5.2.2), (5.2.3), (5.3.2), etc.), and for conics and quadrics we know the general equation (5.2.4). In this final section we want to give some more results of linear and quadratic, projective geometry first. Then, we will briefly consider projective transformations.

5.4.1 Proposition. a) *In a projective plane, and with respect to projective coordinates, three points* $M_a(x_1^a, x_2^a, x_0^a)$, $(a = 1,2,3)$ *are collinear iff*

$$\begin{vmatrix} x_1^1 & x_2^1 & x_0^1 \\ x_1^2 & x_2^2 & x_0^2 \\ x_1^3 & x_2^3 & x_0^3 \end{vmatrix} = 0. \tag{5.4.1}$$

b) In projective space, and with respect to projective coordinates, three noncollinear points M_a $(a = 1,2,3)$ *define a plane of parametric equations*

$$x_i = \lambda_1 x_i^1 + \lambda_2 x_i^2 + \lambda_3 x_i^3, \tag{5.4.2}$$

where $i = 1,2,3,0$, *and* x_i *are the coordinates of any point of the plane. c) Four points* M_a $(a = 1,2,3,4)$ *of space belong to the same plane iff*

$$\begin{vmatrix} x_1^1 & x_2^1 & x_3^1 & x_0^1 \\ x_1^2 & x_2^2 & x_3^2 & x_0^2 \\ x_1^3 & x_2^3 & x_3^3 & x_0^3 \\ x_1^4 & x_2^4 & x_3^4 & x_0^4 \end{vmatrix} = 0. \tag{5.4.3}$$

Proof. a) By (5.3.2), collinearity holds iff the coordinates of M_3 are linear combinations of those of M_1, M_2, and this is equivalent to (5.4.1). (Think of the determinant as a "false mixed product".) b) The result obviously holds for projective coordinates with respect to a frame which has M_1, M_2, M_3 as three of its fundamental points. Then, an arbitrary, projective change of coordinates doesn't change the form of the equations (5.4.2). c) This follows from b), and from classical linear algebra which we do not explain here. Q.E.D.

5.4.2 Proposition. *a) In plane, three straight lines are concurrent iff the determinant of the coefficients of their general equations (5.2.2) is zero. b) In space, four planes have a common point iff the determinant of the coefficients of their general equations (5.2.3) is zero.*

Proof. These conditions are just the plane (space) duals of the conditions (5.4.1), (5.4.3). Q.E.D.

Concerning quadratic geometry, if we consider a conic (quadric) of general equation

$$f(x_i) := \sum_{i,j} a_{ij} x_i x_j = 0, \tag{5.4.4}$$

its intersection with a straight line (5.3.2) consists of the points (5.3.2) where the parameters λ satisfy the equation

$$f(\lambda_0 x_i^0 + \lambda_1 x_i^1) = 0.$$

A straightforward computation shows that this *intersection equation* may be written under the form

$$(\lambda_0)^2 f(x_i^0) + 2\lambda_0\lambda_1 f(x_i^0; x_i^1) + (\lambda_1)^2 f(x_i^1) = 0, \tag{5.4.5}$$

where $f(x_i^0; x_i^1)$ is obtained from f by *polarization* following the rule (3.3.26).

Since the quadratic equation (5.4.5) decomposes into two (possibly complex) linear factors (if we divide the equation by either λ_0^2 or λ_1^2, we get a usual quadratic function of either λ_1/λ_0 or λ_0/λ_1, and we decompose it), we always get two intersection points (with the parameters λ determined up to an arbitrary factor). In particular, no such things as asymptotic directions and asymptotes exist in projective geometry.

If the two intersection points coincide, we will say that the line is *tangent* to the conic (quadric). This happens iff

$$(f(x_i^0; x_i^1))^2 - f(x_i^0) f(x_i^1) = 0. \tag{5.4.6}$$

In particular, if the point (x_0) is fixed, and x_i^1 are replaced by x_i, (5.4.6) is the equation of the tangents to the conic (quadric) through this fixed point. If the latter belongs to the conic (quadric), we remain with a *tangent line (plane)* which has the equation

$$f(x_i^0; x_i) = 0, \tag{5.4.7}$$

just as in affine geometry.

The notion of a *polar line (plane)* of a point can also be defined as in affine geometry (Section 3.3) i.e., as the line (plane) which contains the contact points of the tangents through (x_i^0). Its equation is again (5.4.7), computed by polarization but, (x_i^0) is no longer a point of the conic (quadric). In fact, there exists one more interesting characteristic of the polar line (plane) namely, it is the locus of the points which are harmonically conjugated with the *pole* with respect to the intersection points of the conic (quadric) with lines through the pole (see Problem 5.4.3 at the end of this section).

We will not develop these results in more detail but, to get the flavour, let us solve an exercise.

5.4.3 Exercise. In a projective plane, with a given projective frame A_1, A_2, A_0, U, find the equation of a conic which is tangent to A_1A_0 at A_1, is tangent to A_2A_0 at A_2, and it passes through the unit point U.

Solution. The conic has a general equation (5.4.4), and, if we ask $A_1(1,0,0)$ and $A_2(0,1,0)$ to satisfy this general equation we get

$$a_{11} = 0, \quad a_{22} = 0.$$

Then, we polarize the remaining equation at A_1, which yields

$$a_{12}x_2 + a_{10}x_0 = 0,$$

and ask the result to be the line A_1A_0 i.e., $x_2 = 0$. This shows that $a_{10} = 0$. The similar operation at A_2 yields $a_{20} = 0$. Therefore, the equation of the conic must be

$$2a_{12}x_1x_2 + a_{00}x_0^2 = 0,$$

and we must still ask $U(1,1,1)$ to satisfy this equation i.e., $2a_{12} = -a_{00}$. Accordingly, the equation of the required conic is

$$x_1x_2 - x_0^2 = 0. \tag{5.4.8}$$

Furthermore, it is worthwhile to briefly indicate the projective classification of conics and quadrics in real projective geometry, and this also can be achieved easily.

Namely, the same Gauss method, which we used to get the affine classification theorems 3.4.5, 3.4.6, allows us to put the projective equation (5.4.4) of a conic (quadric) in the *canonical form* of an algebraic sum of squares. If we think of all the possible sums of squares, we get

5.4.4 Theorem. *In plane, real, projective geometry, the conics are classified as follows:*
1) Nondegenerate Imaginary Conics, of canonical equation

$$x_1^2 + x_2^2 + x_0^2 = 0.$$

(There are no real points on such a conic.)
2) Nondegenerate Real Conics, of canonical equation

$$x_1^2 + x_2^2 - x_0^2 = 0.$$

3) Pairs of imaginary and real straight lines:

$$x_1^2 \pm x_2^2 = 0.$$

4) Two coincident lines: $x_1^2 = 0$. *(In cases 3 and 4, the conic is degenerate.)*

Here, the main fact is that there is a single type of real, nondegenerate conics (class 2), and this means that, from the projective point of view, ellipses, hyperbolas and parabolas are equivalent!

Similarly, in projective space we have

5.4.5 Theorem. *In real, projective space the nondegenerate quadrics are classified as follows:*
1) Imaginary Quadrics, of canonical equation

$$x_1^2 + x_2^2 + x_3^2 + x_0^2 = 0.$$

(There are no real points on such a quadric.)
2) Quadrics of Elliptic Type, of canonical equation

$$x_1^2 + x_2^2 + x_3^2 - x_0^2 = 0.$$

3) Quadrics of Hyperbolic Type, of canonical equation

$$x_1^2 + x_2^2 - x_3^2 - x_0^2 = 0.$$

The other (degenerate) quadrics are either cones or pairs of planes.

5.4.6 Example. Establish the canonical equation, and the type of the conic (5.4.8) encountered in Exercise 5.4.3.

Solution. The projective change of coordinates

$$x_1 = \frac{1}{2}(y_1 + y_2),\ x_2 = \frac{1}{2}(y_1 - y_2),\ x_0 = \frac{1}{2}y_0$$

replaces (5.4.8) by the canonical equation

$$y_1^2 - y_2^2 - y_0^2 = 0,$$

and this is a nondegenerate real conic.

5.4.7 Remark. *As in affine geometry (Proposition 3.4.8), we can see that the annulation of $\Delta := det(a_{ij})$, where a_{ij} are the coefficients of the general equation of the conic (quadric) is invariant by projective coordinate transformations, and that the conic (quadric) is nondegenerate iff $\Delta \neq 0$.*

Now, following again the path of affine geometry (see Section 4.2), we can define and study projective transformations.

5.4.8 Definition. *Let Π, Π' be two models of projective lines (planes, spaces), and let $\pi : \Pi \longrightarrow \Pi'$ be a geometric transformation (Definition 4.1.1). Then, π is called a **projective transformation** if there exist projective coordinate systems of Π, Π', respectively, such that the coordinates of two corresponding points are equal. (These coordinate systems, and their associated frames, **correspond** each to the other by π.) Furthermore, if the elements of Π, Π' have the same geometric role (e.g., points), the projective transformation π is a **collineation** or a **homography**, and, if the elements of Π, Π' have the dual roles (e.g., points and lines or planes), π is a **correlation**.*

The following results have exactly the same proof as in affine geometry (Section 4.2)

5.4.9 Theorem. a) *For every pair of projective frames (A_i, U), (A'_i, U'), in Π, Π', respectively, there exists one and only one projective transformation which makes the two frames correspond, and this transformation has the equations*

$$\rho x'_i = x_i \qquad (0 \neq \rho \in \mathbf{R}), \tag{5.4.9}$$

where x_i (x'_i) are projective coordinates associated to the two frames, respectively.
b) If $\pi : \Pi \longrightarrow \Pi'$ is a projective transformation, its equations with respect to arbitrary frames of the two spaces are of the form

$$\rho x'_i = \sum_j a_{ij} x_j, \tag{5.4.10}$$

where ρ is an arbitrary, nonzero factor, and $det(a_{ij}) \neq 0$.
c) A projective transformation preserves (or dualizes, if a correlation) collinearity and coplanarity of points, as well as the absence of these properties, and it sends lines onto lines and planes onto planes (if it is a collineation).
d) A projective transformation preserves the cross ratios.

5.4.10 Remark. *One can prove that a geometric transformation between projective planes or spaces is projective iff it preserves or dualizes collinearity. To get a projective transformation between two lines one must also ask for the preservation of the cross ratios.*

One of the most interesting situations is that of projective transformations in a fixed projective space, $\pi : \Pi \longrightarrow \Pi$. It is easy to understand that the set of all these transformations is a transformation group $\mathcal{P}(\Pi)$, called the *projective group* of Π, and the *projective geometry* of Π is the study of the invariants and invariant properties of this group. We may also notice that, if an improper plane α is chosen in the space Π (or an improper line, if Π is a plane), the projective transformations of Π which preserve α may be identified with the affine transformations of the affine space $\Pi \setminus \alpha$. In this sense, the affine group may be seen as a subgroup of the projective group, and the projective invariants are also affine invariants. But the converse is not true. For instance, an ellipse and a hyperbola or a parabola are projectively equal (there exists a projective transformation which sends the former into the latter) but, they are different from the affine point of view.

If $\pi : \Pi \longrightarrow \Pi$ is a projective transformation, the points, lines or planes which remain unchanged by π are said to be *fixed*, and they characterize π, in many respects. Let us indicate how to look for the fixed points of such a transformation. With respect to a chosen projective frame, π has the equations (5.4.10), where, say, $M(x_i)$, and $M'(x'_i)$

are corresponding points. Since these equations already contain the arbitrary factor ρ, we may say that M is a fixed point if $x'_i = x_i$, and, then, (5.4.9) becomes

(5.4.11) $$\sum_j (a_{ij} - \rho\delta_{ij})x_j = 0$$

($\delta_{ij} = 1$ if $i = j$, $\delta_{ij} = 0$ if $i \neq j$). Moreover, we need nontrivial solutions of (5.4.11) (i.e., not all $x_i = 0$). Therefore, here, we have the same algebraic problem as in the case of the symmetry axes of a conic (quadric) (e.g., (3.3.36), (3.3.37)). Hence, first, we must solve the *characteristic equation*

(5.4.12) $$det(a_{ij} - \rho\delta_{ij}) = 0,$$

then insert the solutions into the *characteristic system* (5.4.11), and solve it to get the required fixed points.

For instance, if Π is a projective line, (5.4.9) is

(5.4.13) $$\rho x'_1 = a_{11}x_1 + a_{10}x_0,$$

$$\rho x'_0 = a_{01}x_1 + a_{00}x_0.$$

The corresponding characteristic equation (5.4.12) is a quadratic equation, and it has two solutions which may be: complex (*the elliptic case*), real and distinct (*the hyperbolic case*), real and equal (*the parabolic case*). In the hyperbolic case, if we use the two fixed points as fundamental points of a projective frame, the equations of the transformation will appear under the *canonical form*

(5.4.14) $$\rho x'_1 = \rho_1 x_1, \ \rho x'_0 = \rho_2 x_0,$$

etc.

5.4.11 Remark. *In the study of the projective transformations between lines, it is sometimes useful to represent the transformation by an equation in nonhomogeneous coordinates. Namely, if we divide the first equation (5.4.13) by the second equation, and put $x' = x'_1/x'_0$, $x = x_1/x_0$, the equation of the transformation becomes*

(5.4.15) $$x' = \frac{ax + b}{cx + d},$$

where a, b, c, d are some coefficients which satisfy the condition $ad - bc \neq 0$. (5.4.15) is the "classical" form of the homographic transformations of a line.

Exercises and Problems

5.4.1. In a projective plane, show that the conic

$$(\Gamma) \quad x_1 x_2 + x_2 x_0 + x_0 x_1 = 0$$

passes through the fundamental points A_1, A_2, A_0 of the projective frame. Show that the line which joins A_1 with the intersection point of the tangents to Γ at A_2, A_0 contains the unit point of the frame. What is the projective type of Γ?

5.4.2. A conic Γ passes through the vertices of ΔABC. Prove that the intersection points of the sides of ΔABC with the tangent at the opposite vertex are collinear.

5.4.3. Prove that the polar line (plane) of a point M with respect to a conic (quadric) Γ is the locus of the point N which is harmonically conjugated to M with respect to the intersection points of Γ with the straight lines through M.

5.4.4. Prove that the polar lines of three collinear points with respect to a conic are concurrent lines.

5.4.5. Find the canonical equation and the projective type of the following conics and quadrics

$$2x_1^2 + x_2^2 - x_0^2 + 3x_1 x_2 - x_1 x_0 + 2x_2 x_0 = 0,$$

$$5x_1^2 + x_2^2 + x_0^2 + 2x_1 x_2 - 4x_1 x_0 = 0,$$

$$x_1^2 + x_3^2 - 2x_3 x_0 + 4x_1 x_2 = 0,$$

$$4x_1^2 + x_2^2 + 5x_3^2 + 4x_1 x_2 - 12x_1 x_3 - 6x_2 x_3 + 3x_1 x_0 - x_2 x_0 = 0,$$

$$x_1^2 + 4x_2^2 + x_3^2 - 2x_0^2 - 4x_1 x_2 + 2x_1 x_3 - 4x_2 x_3 - 6x_3 x_0 = 0.$$

5.4.6. Let $A_1, A_2, A_3, A_4, A_5, A_6$ be six points of a nondegenerate conic Γ. Prove that

$$(A_1A_3, A_1A_4; A_1A_5, A_1A_6) = (A_2A_3, A_2A_4; A_2A_5, A_2A_6).$$

Conversely, the conic Γ through A_1, A_2, A_3, A_4, A_5, where no three of the five points are collinear, is the locus of the intersection point of the lines A_1A_6, A_2A_6 of the pencils with vertices A_1, A_2 which satisfy the previous equality of cross ratios (Steiner's Theorem).

5.4.7. Let $A_1A_2A_3A_4A_5A_6$ be a hexagon with vertices on a conic Γ. Prove that the intersection points of the three pairs of opposite sides of the hexagon are collinear (Pascal's Theorem).

5.4.8. In projective space, consider the points

$$A[1:-1:0:0],\ B[0:1:-1:0],\ C[0:0:1:-1],\ D[1:0:0:1].$$

i)Show that they are the vertices of a tetrahedron, and write down the equations of its face planes and edge lines. ii)Write down the equations of the tangent planes of the quadric

$$(\Gamma) \qquad x_1^2 + x_2^2 - x_3^2 - x_0^2 = 0$$

at those of the vertices of the tetrahedron which belong to Γ. iii) Find the locus of the poles with respect to Γ of the planes which pass through those vertices of the tetrahedron which do not belong to Γ.

5.4.9. Consider the projective transformation

$$\begin{aligned} \rho x_1' &= x_1 - 2x_2 + 3x_0 \\ \rho x_2' &= 2x_1 + x_2 + 2x_0 \\ \rho x_0' &= 4x_1 - 2x_2 + 5x_0 \end{aligned}$$

of an enlarged affine plane. Find an improper point of the plane which is the image of another improper point by this transformation.

5.4.10. Find the equations of the projective transformation of a projective plane which sends the point of coordinates $[1:0:0]$ to $[0:1:0]$, the point $[0:1:0]$ to $[0:0:1]$, the point $[0:0:1]$ to $[1:0:0]$, and $[1:1:1]$ to $[1:2:4]$. Find the fixed points of this transformation.

5.4.11. Find the fixed points of the projective transformation of space defined by

$$\rho x_1' = x_3,\ \rho x_2' = 2x_0,\ \rho x_3' = 4x_2,\ \rho x_0' = x_1 + x_2 \quad (\rho \neq 0).$$

5.4.12. A plane projective transformation is called a *hyperbolic homology* if it has a fixed point O (*center*) and a line d of fixed points (*axis*) which does not contain O. Prove that the transformation

$$x_1' = x_2 + x_0,\ x_2' = x_1 + x_0,\ x_0' = x_1 + x_2$$

is a hyperbolic homology, and find its center and axis.

5.4.13. Let φ be a hyperbolic homology of the plane with center O and axis d. φ is said to be *involutive* if $\varphi^{-1} = \varphi$. If this happens, and if M, $M' = \varphi(M)$, for M not on d, and $K = OM \cap d$, show that O, K, M, M' are collinear points, and $(O, K; M, M') = -1$.

5.4.14. Extend the notion of a homology, and the result of Problem 5.4.13 to projective space, by replacing the axis by a plane.

Appendix

Analytical Geometry via Maple

In this Appendix, we want to indicate a possibility of using a computer in doing computations relevant to problems of analytical and projective geometry. Our choice is Maple V, an interactive computer algebra system, which is both simple, and accessible, even with a limited computer equipment. In particular, Maple can be installed on the IBM compatible personal computers now in use.

We do not intend to teach Maple here, interested reader should take care of acquiring the necessary hardware and software equipment, and learn how to use it from the enclosed documentation. The only thing we will do is to direct the reader to those parts of Maple which are useful in studying analytical and projective geometry. We do this following the First Leaves [4], and the Library Reference Manual of Maple V [6]. (See also the Language Reference Manual [5].) As a matter of fact, for help concerning the various commands, one can either ask for on-line help during the work with the computer or consult the above mentioned manuals.

The various operations of vector algebra are covered by the Linear Algebra Package of the Maple Library. This package is also useful for other purposes of analytical geometry. Hence, at the beginning of a work session of solving problems of analytical geometry, after invoking Maple by the command

> *maple*

(or its equivalents; > denotes the prompt), we will ask

$> with(linalg):$

Do not forget to press the RETURN key after each command.

Then, if we assume that a coordinate system already exists (this system being either affine or orthogonal, according to the necessities which we know, but the computer does not), we may define vectors as in the following example

$> v := vector([a, b, c]);$

In accordance with the Maple syntax, here, v is the name of the vector, and (a, b, c) are the coordinates of the vector. They may be either numbers or algebraic symbols since Maple does symbolic computations too. Let us also recall that a Maple command ends either with a colon : and then it is executed but, the result is not displayed, or by a semicolon ; and then the result is displayed. It is also worthwhile to recall that the vectors are, in fact, a particular case of matrices. We recall one of the possibilities to define an m by n matrix A (i.e., with m rows and n columns):

$> A := matrix(m, n, [a(1,1), a(1,2), \ldots, a(m,n)]);$

where $a(i, j)$ is the element in row i and column j, and first we write all the elements of the first row, then all the elements of the second row, and so on.

A sum of vectors will be calculated as follows

$> v := vector([a1, b1, c1]):$

$> w := vector([a2, b2, c2]):$

$> add(v, w);$

and the result will be

$> [a1 + a2, b1 + b2, c1 + c2]$

Another possibility is

$> evalm(v + w);$

with the same result.

The multiplication of the vector v above by a scalar $\lambda = lambda$ (so denoted because we do not have Greek letters in Maple), is performed by the command

> $evalm(lambda * v);$

In the same way, one computes sums of matrices of the same type, and the product of a matrix by a scalar. But, the multiplication of two matrices A, B of type (m, n), (n, p), respectively, is given by

> $evalm(A\&* B);$

or by

> $multiply(A, B);$

The scalar product of two vectors, say v and w defined above, may be computed either by the command

> $dotprod(v, w);$

or by the command

> $innerprod(v, w);$

There are also commands for the calculation of the length of a vector

> $norm(v, frobenius);$

and of the angle of two vectors

> $angle(v, w);$

The Linear Algebra Package computes vector products, by

> $crossprod(v, w);$

and mixed products, seen as determinants by

> $A := matrix(3, 3, [u1, u2, u3, v1, v2, v3, w1, w2, w3]) : det(A);$

where the brackets contain the coordinates of the three vectors u, v, w whose mixed product we want to compute.

More complicated operations will be performed by decomposing them into simpler operations, and by using the interactive character of Maple. On the other hand, other functions of the Linear Algebra Package are useful for more complicated questions of geometry. For instance, *eigenvals* and *eigenvects* ([6], pp. 370-371) are essential in finding the axes of a conic or quadric (see Section 3.3). The commands *linsolve* [6], p. 396, which solves systems of linear equations, and $inverse(A)$; or $evalm(1/A)$; which compute the inverse of a non-singular, square matrix A, are useful in various instances.

The Maple Library has three specifically geometric packages: the Euclidean Geometry Package, invoked by

$> with(geometry):$

the 3-D (i.e., three dimensional) Geometry Package, invoked by

$> with(geom3d):$

and the Projective Geometry Package, invoked by

$> with(projgeom):$

The first and the third of these packages are concerned with problems of plane geometry only. All three packages contain a lot of interesting geometric functions, and may be enlarged even more by composition of functions which uses interactive work too. We notice that some of these functions yield an answer of *true* or *false* for some specified questions. This is possible since Maple makes boolean computations. We will refer to a small number of examples for each of these packages.

In the Euclidean Geometry Package, we may define a point by

$> point(A, a1, a2);$ or $> point(A, [a1, a2]);$

where $a1, a2$ are the coordinates of the point A.

If we define three points A, B, C in this way, the command

$> are\_collinear(A, B, C);$

will give us the answer *true* or *false*, as is the case.

On the other hand, if we ask

$> triangle(ABC, [point(A, a1, a2), point(B, b1, b2), point(C, c1, c2)]):$

$area(ABC);$

we get the area of the triangle in return.

In the Euclidean Geometry Package, there are commands which test concurrence, parallelism and orthogonality of straight lines in plane as in the following example

$> line(l1, [a*x+b*y+c=0]), line(l2, [e*x+f*y+g=0]):$

$> are\_parallel(l1, l2);$

The result will be *true* or *false* as is the case.

Other things which we can compute are the distance between two points and the distance from a point to a straight line. For instance,

$> point(A, a1, a2) : line(l, [a*x+b*y+c=0]) : distance(A, l);$

And, for the same point A and line l, the commands

$> perpendicular(A, l, lp) : lp[equation];$

provide us with the equation of the perpendicular line from A to l, a line called here lp.

Among other things which the functions of the package can do are: define conics of various types, compute polars and poles, compute radical axes and centers of circles, compute symmetric points, compute tangent lines, etc. [6].

The 3-D Geometry Package also has a large number of functions, and it may be successfully used in three dimensional geometric computations. With this package, a point A is defined as follows

$> point3d(A, a1, a2, a3);$ or $> point3d(A, [a1, a2, a3]);$

where $(a1, a2, a3)$ are the coordinates of the point.

If we give two points A, B, the command

$> line3d(l, [A, B]) : l[equation];$

provides the parametric equations of l. Of course, if we give a point A and a direction vector v, we can get a second point immediately, and define the line in Maple again.

A plane by three points A, B, C may be defined by the command

$> plane(p, [A, B, C]) : p[equation];$

If the content of the bracket of this command is replaced by either a point and two vectors or a linear equation or polynomial in x, y, z, the command will again define the corresponding plane. There are many other possibilities of writing down equations of planes and lines. For instance, a plane or a line through a given point and either parallel or perpendicular to a given plane or line, etc.

The commands of the form

$> angle(U, V);$

where U and V are either lines defined by pairs of points, or planes defined by linear equations, or a line and a plane, compute the angle of the mentioned objects.

There are commands to compute the distance between two points, a point and a line, two lines, and some areas and volumes in space.

The package also covers the usual geometry of spheres: definition, center, radius, tangent planes, power of a point, etc.

There are no specific commands to deal with quadrics, but one can solve problems concerning quadrics by an interactive work with the computer, and by a convenient use of the Linear Algebra Package. We already mentioned the search of axes via the eigenvalues and eigenvectors of the operator T of the quadric (Section 3.3). Intersections with a line, and tangency questions may be answered by solving the system of equations using Maple. Classification problems may be answered by computing the invariants δ, Δ, etc. (See an example later on in this Appendix.)

Finally, let us mention some of the possibilities of the Projective Geometry Package. As mentioned, the package deals with plane projective geometry. It should not be used together with the Euclidean Geometry Package in order to avoid confusions.

Points are defined by projective, homogeneous coordinates as follows

$> point(A, [a1, a2, a0]);$

where A is the name of the point. Straight lines are defined by a similar 3-sequence, where we have the coefficients of the homogeneous equation of the line instead of coordinates. Conics are introduced by the six coefficients of their homogeneous equation e.g.,

$> conic(C, [a11, a22, a00, 2a20, 2a10, 2a12]);$

the order of the coefficients being as indicated, and in our usual notation for the general equation of a conic (Sections 3.3, 5.4).

There are commands to check collinearity of points and coplanarity of lines, to compute harmonic conjugated points with respect to a given pair, to compute intersection points of a conic with a line, to write equations of tangents and polars (poles) with respect to a conic, etc. [6].

We notice that it is possible to extend the use of Maple to solve questions related to projective transformations, and space projective geometry by an adequate use of the Linear Algebra Package.

We end by examples of more involved, interactive sessions of Maple which answer analytical geometry problems.

Problem 1. Let $l1$ be the straight line in space which passes through the points $A(1,0,-1)$, $B(-1,2,1)$ and $l2$ the line defined by the equations

$$2x + 3y - z + 1 = 0, \quad x - y + z - 2 = 0.$$

Show that these are skew lines, compute the distance between them, and write down the equations of their common perpendicular (Section 2.3).

Solution. We begin by introducing our data in the computer

$> maple$

$> with(linalg):$

$> with(geom3d):$

$> point3d(A, [1, 0, -1]);$

$> point3d(B, [-1, 2, 1]);$

$> line3d(l1, [A, B]);$

Furthermore (see Section 2.3), we need a point and a direction vector for $l1$ and $l2$. On $l1$, we may use the point A and the vector $v1 = \vec{AB}$

$> v1 := vector([-2, 2, 2]);$

In order to find a point of $l2$, we ask

$> solve(\{2*x+3*y-z+1=0, x-y+z-2=0\}, \{x,y\});$

This asks for solutions of the system of equations of $l2$ with respect to x, y, if z is seen as a parameter. If no solutions would be found, we would ask for solutions with respect to either (y, z) or (x, z). But, in our case, on the screen we get the answer

$$\{y=(3/5)z-1, \quad x=-(2/5)z+1\},$$

hence, we may use the point

$> point3d(M, [1, -1, 0]);$

Then, the vector product of the normal vectors of the two planes which have $l2$ as intersection line is a vector $v2$ of $l2$:

$> v2 := crossprod([2, 3, -1], [1, -1, 1]);$

with the result

$$v2 := [2, -3, -5]$$

The corresponding second point of $l2$ is

$> point3d(N, [3, -4, -5]);$

and then we give the command

$> line3d(l2, [M, N]);$

The following command

$> plane(p, [A, B, M]), on\_plane(N, p);$

checks that the lines are skew because the answer received on the screen is *false*.

The next command gives us the distance between the two lines

$> distance(l1, l2);$

The answer is $(1/7)56^{1/2}$.

We also need the vector product of $v1$ and $v2$ (Section 2.3)

$> v3 := crossprod(v1, v2);$

with the result

$$v3 := [-4, -6, 2]$$

and the generic radius vector

$> r := vector([x, y, z]);$

Then, the common perpendicular is defined by the equations

$> det(matrix(3, 3, [x - 1, y, z + 1, -2, 2, 2, -4, -6, 2])) = 0,$
$> det(matrix(3, 3, [x - 1, y + 1, z, 2, -3, 5, -4, -6, 2])) = 0;$

with the result

$$16x + 4 - 4y + 20z = 0, \quad 24x - 8 + 16y = 0.$$

Problem 2. Find the type, and the affine canonical equation of the quadric

$$x^2 + 2y^2 - 3z^2 + xy - 3yz + 4zx - 2x + y - z - 10 = 0.$$

Solution. Of course, we can go directly to the canonical equation, and, thereby, find the type of the quadric too. But, in order to exercise more, we will start by establishing the type first:

$> maple$

$> with(linalg):$

$> a := matrix(3, 3, [1, 1/2, 2, 1/2, 2, -3/2, 2, -3/2, -3]):$

$> A := matrix(4, 4, [1, 1/2, 2, -1, 1/2, 2, -3/2, 1/2, 2, -3/2, -3, -1/2,$
$-1, 1/2, -1/2, -10]):$

$> delta := det(a);$

$> Delta := det(A);$

On the screen, we will see the results *delta* $:= -37/2$, *Delta* $:= 3185/16$ (i.e., in the notation of Section 3.4, $\delta := -37/2$, $\Delta := 3185/16$), and the classification tables of Section 3.4 tell us that the quadric is either an imaginary ellipsoid or a one-sheeted hyperboloid. In order to decide, we consider the intersection of the cone of asymptotic directions with vertex at the origin

$$x^2 + 2y^2 - 3z^2 + xy - 3yz + 4zx = 0$$

with planes $z = k = const.$, and we will have to say what is the type of the conics

$$x^2 + 2y^2 + xy + 4kx - 3ky - 3k^2 = 0.$$

For this purpose, we ask

$> b := matrix(2, 2, [1, 1/2, 1/2, 2]);$

$> B := matrix(3, 3, [1, 1/2, 2*k, 1/2, 2-(3*k)/2, 2*k, -(3*k/2), -3*k^2]);$

$> det(b);$

$> det(B);$

On the screen, the answers are 7/4 for $det(b)$, and $-(37/2)k^2$ for $det(B)$. Hence, for $k = 0$ the conic is an imaginary pair of crossing lines, and for $k \neq 0$ the conic is an ellipse. To say whether this ellipse is real or imaginary, we cut it with the lines $y = t = const.$. The intersection equation will be

$$x^2 + (t + 4k)x + (2t^2 - 3tk - 3k^2) = 0,$$

and its discriminant is introduced in Maple by

$> D := simplify((t + 4*k)^2 - 4*(2*t^2 - 3*t*k - 3*k^2));$

We are given the result

$$D := -7t^2 + 20tk + 28k^2,$$

and, since $k \neq 0$, we may write

$> u := t/k : D := k^2 * (-7*u^2 + 20*u + 28);$

Then, we will ask

> $solve(-7 * u^2 + 20 * u + 28 > 0, u);$

and get the answer on the screen:

$$\{u < 10/7 + (2/7)74^{1/2},\ 10/7 - (2/7)74^{1/2} < u\}$$

Therefore, there are values of t, k which give a real ellipse, the asymptotic cone must be real, and the quadric which we study is a one-sheeted hyperboloid.

Now, let us find the affine canonical equation of our quadric by the Gauss method. First, we give a Maple name to the left hand side of the equation of the quadric

> $Q := x^2 + 2y^2 - 3z^2 + xy - 3yz + 4zx - 2x + y - z - 10;$

Then, we may proceed by

> $collect(Q, x);$ a command which gives the result on the screen

$$x^2 + (-2 + y + 4z)x - 3yz + 2y^2 - 3z^2 + y - z - 10.$$

This shows that the first "square" of the Gauss method should be here:

> $Q1 := (x + (1/2) * (-2 + y + 4 * z))^2 :$

Now, we are going to see what the remaining part of the quadric's equation is:

> $R1 := simplify(Q - Q1);$

The result which appears on the screen is

$$R1 := (7/4)y^2 - 7z^2 - 5yz + 2y + 3z - 11.$$

The continuation of the process is similar, and we write the necessary Maple commands without further comments.

> $collect(R1, y);$

Answer:

$$(7/4)y^2 + (2 - 5z)y + 3z - 7z^2 - 11.$$

$> Q2 := (7/4) * (y + (2/7) * (2 - 5 * z))^2 :$

$> R2 := simplify(R1 - Q2);$

Answer:

$$R2 := -(74/7)z^2 + (41/7)z - 81/7.$$

$> Q3 := -(74/7) * (z - (41/148))^2 :$

$> CQ := Q1 + Q2 + Q3 + simplify(R2 - Q3);$

CQ is the left hand side of the equation of the quadric written as a "sum of squares", and the result on the screen is

$$CQ := (x - 1 + (1/2)y + 2z)^2 + (7/4)(y + 4/7 - (10/7)z)^2$$

$$-(74/7)(z - 41/148)^2 - 3185/296.$$

This result shows that the quadric is a one-sheeted hyperboloid, and, also, it shows what affine change of coordinates should be performed in order to obtain the affine canonical equation of the quadric. Namely:

$$\tilde{x} = x + \frac{1}{2}y + 2z - 1,$$

$$\tilde{y} = \sqrt{\frac{7}{4}}(y - \frac{10}{7} + \frac{4}{7}),$$

$$\tilde{z} = \sqrt{\frac{74}{7}}(z - \frac{41}{148}).$$

Problem 3. Find the space projective transformation which sends the points $[-5 : 7 : 14 : -8]$, $[2 : -1 : 3 : 9]$, $[-4 : 3 : 2 : -1]$, $[1 : 7 : 9 : 11]$, $[-5 : 3 : 1 : 2]$ to $[3 : 5 : 7 : 9]$, $[2 : 4 : 11 : 20]$, $[-1 : -6 : 5 : 17]$, $[7 : 11 : -13 : 15]$, $[24 : -15 : -9 : 8]$, respectively.

Solution. The homogeneous coordinates of a point may be seen as a one-column matrix, and those of four points yield a corresponding 4 by 4 matrix, say $V1$ in our case. Then, if the equations of a space projective transformation, described in Section 5.4, are used to compute the images of four given points, the results contain four different arbitrary factors, and may be arranged as four columns of a matrix, say $V2$. The coefficients of the transformation form again a 4 by 4 matrix, say A, and the transformation just means $V2 = A(V1)$, where the right hand side is a product of two matrices. Accordingly, we begin our Maple session by the following commands

$> maple$

$> with(linalg):$

$> V1 := matrix(4,4,[-5,7,14,-8,2,-1,3,9,-4,3,2,-1,1,7,9,11]);$

$> V2 := matrix(4,4,[3*e,5*e,7*e,9*e,2*f,4*f,11*f,20*f,$
$-g,-6*g,5*g,17*g,7*h,11*h,-13*h,15*h]);$

(e, f, g, h are the arbitrary factors due to the homogeneity of the coordinates)

$> A := multiply(V2, inverse(V1));$

On the screen, we get the matrix

$$\begin{pmatrix} \frac{96}{3001}e & -\frac{556}{3001}e & -\frac{1948}{3001}e & \frac{2803}{3001}e \\ \frac{26}{3001}f & \frac{4601}{3001}f & \frac{1223}{3001}f & \frac{1822}{3001}f \\ \frac{297}{3001}g & \frac{10659}{3001}g & \frac{4852}{3001}g & -\frac{3426}{3001}g \\ -\frac{6144}{3001}h & -\frac{12432}{3001}h & -\frac{1370}{3001}h & \frac{9671}{3001}h \end{pmatrix}$$

Now, we look at the fifth pair of the corresponding points:

$> u1 := vector([-5,3,1,2]);$

$> u2 := vector([24*t,-15*t,-9*t,8*t]);$

where we introduced a homogeneity factor again, and compute

$> multiply(A, u1);$

The result given by the computer is

$$[\frac{1510}{3001}e, \frac{18540}{3001}f, \frac{28492}{3001}g, \frac{11396}{3001}h].$$

Since the result must be $u2$, we get (without the computer)

$$e = \frac{3001 * 24}{1510} t,\ f = -\frac{3001 * 15}{18540} t,\ g = -\frac{3001 * 9}{28492} t,\ h = \frac{3001 * 8}{11396} t.$$

The substitution of these values in the matrix A, gives us the matrix of the required transformation, and we are done. (Of course, the number multiplications involved will be done by the computer again.)

Solutions of Exercises and Problems

1.2.1. i) We look at

$$\lambda\bar{l}+\mu\bar{m}+\nu\bar{n}=\bar{a}(-\lambda+2\mu-\nu)$$
$$+\bar{b}(2\lambda-\mu-\nu)+\bar{c}(-\lambda-\mu+2\nu)=\bar{0}.$$

Since $\bar{a}$, $\bar{b}$, $\bar{c}$ are independent, we must have

$$-\lambda+2\mu-\nu=0,\ 2\lambda-\mu-\nu=0,\ -\lambda-\mu+2\nu=0,$$

and this holds for $\lambda=\mu=\nu$, possibly nonzero. Hence, $\bar{l}$, $\bar{m}$, $\bar{n}$ are linearly dependent vectors.

ii) Look for coefficients λ, μ, ν such that

$$\bar{s}=\lambda\bar{l}'+\mu\bar{m}'+\nu\bar{n}'.$$

Since $\bar{a}$, $\bar{b}$, $\bar{c}$ are independent, we get $\lambda=4/5$, $\mu=1/5$, $\nu=3/5$.

1.2.2. If O is an arbitrary origin, we have

$$\vec{EF}=\frac{1}{2}(\bar{r}_C+\bar{r}_D)-\frac{1}{2}(\bar{r}_B+\bar{r}_A)$$
$$=\frac{1}{2}[(\bar{r}_D-\bar{r}_A)]+\frac{1}{2}[(\bar{r}_C-\bar{r}_B)]=\frac{1}{2}(\vec{AD}+\vec{BC}).$$

Similarly,

$$\vec{MN}=\frac{1}{2}(\vec{AB}+\vec{DC}),$$

and

$$\vec{MN}+\vec{EF}=\frac{1}{2}(\vec{AB}+\vec{DC}+\vec{AD}+\vec{BC})=\vec{AB}+\vec{BC}=\vec{AC}.$$

1.2.3. Use Fig. 1.2.8. The first assertion follows from

$$\vec{AA'} + \vec{BB'} + \vec{CC'} = \frac{1}{2}(\vec{AB} + \vec{AC} + \vec{BC} + \vec{BA} + \vec{CA} + \vec{CB}) = \bar{0}.$$

Now the median, say $\bar{m}_A$, of the side $\vec{AA'}$ of the new triangle will be

$$\bar{m}_A = \frac{1}{2}(\vec{CC'} - \vec{BB'}) = \frac{1}{4}(\vec{CA} + \vec{CB} - \vec{BA} - \vec{BC}) = \frac{3}{4}\vec{BC},$$

and similar results hold for the two other medians.

Hence, the sides of the second triangle obtained by the indicated construction are proportional to those of the initial triangle, and the ratio is 3/4.

1.2.4. Use the same method as in the Solved Problem 1.2.21 while taking the vertex A of the tetrahedron $ABCD$ as origin, and joining A with the centroid G_A of ΔBCD, etc. One finds that the centroid G of the tetrahedron satisfies $(A, G_A; G) = 3/4$. With respect to an arbitrary origin O one has

$$\bar{r}_G = \frac{1}{4}(\bar{r}_A + \bar{r}_B + \bar{r}_C + \bar{r}_D).$$

1.2.5. Since $M \in \text{plane}(A'BC)$, we have

$$\vec{A'M} = \lambda\vec{A'B} + \mu\vec{A'C},$$

and if we take S as origin we get

$$\bar{r}_M = \bar{r}_{A'} + \lambda(\bar{r}_B - \bar{r}_{A'}) + \mu(\bar{r}_C - \bar{r}_{A'}).$$

But $\bar{r}_{A'} = \alpha\bar{r}_A$. Thus $\bar{r}_M$ becomes a linear combination of $\bar{r}_A$, $\bar{r}_B$, $\bar{r}_C$.

Then, using $M \in \text{plane}(B'CA)$, $M \in \text{plane}(C'AB)$, we get two more such expressions of $\bar{r}_M$ and the equality of the three expressions of $\bar{r}_M$ yields relations among α, λ, $\mu, \ldots$. A similar procedure will be applied to N. The final results show that $\bar{r}_M$ and $\bar{r}_N$ are proportional.

1.3.1. A is in octant IV and B in octant VI of Fig. 1.3.2.

1.3.2. If we reflect in the plane xOy we get $P'(3, -1, -2)$, $M'(a, b, -c)$.

Reflection in Ox yields $P''(3,1,2)$, $M''(a,-b,-c)$. Reflection in the origin yields $P'''(-3,1,2)$, $M'''(-a,-b,-c)$, etc.

1.3.3. We have $A(0,0)$, $\vec{AB} = \bar{e}_1$, hence $B(1,0)$, $\vec{AC} = \bar{e}_2$ hence $C(0,1)$, $\vec{AD} = 2\vec{BC} = 2(\bar{e}_2 - \bar{e}_1)$ hence $D(-2,2)$, $\vec{AE} = \vec{AD} + \vec{DE} = -2\bar{e}_1 + 2\bar{e}_2 - \bar{e}_1$ hence $E(-3,2)$, and $\vec{AF} = \vec{CD} = \bar{r}_D - \bar{r}_C = -2\bar{e}_1 + 2\bar{e}_2 - \bar{e}_2$ hence $F(-2,1)$.

1.3.4. $(\pm a/\sqrt{2},0,0)$, $(0,\pm a/\sqrt{2},0)$, $(\pm a/\sqrt{2},0,a)$, $(0,\pm a/\sqrt{2},a)$.

1.3.5. We may write $\bar{i} = \vec{AB}$, $\bar{j} = \vec{AD}$, $\bar{i}' = \vec{OB}$, $\bar{j}' = \vec{OC}$, and we get

$$\bar{i}' = \frac{1}{2}(\bar{i} - \bar{j}),\ \bar{j}' = \frac{1}{2}(\bar{i} + \bar{j}), \vec{AO} = \frac{1}{2}(\bar{i} + \bar{j}).$$

Then, an equality $\vec{AM} = \vec{AO} + \vec{OM}$ yields

$$x\bar{i} + y\bar{j} = \frac{1}{2}(\bar{i} + \bar{j}) + \frac{x'}{2}(\bar{i} - \bar{j}) + \frac{y'}{2}(\bar{i} + \bar{j}),$$

and the result is

$$x = \frac{1}{2}(x' + y' + 1),\ \ y = \frac{1}{2}(-x' + y' + 1).$$

1.3.6. Let A, B, C be the common endpoints of $(\bar{i}, \bar{i}')$, $(\bar{j}, \bar{j}')$, $(\bar{k}, \bar{k}')$, respectively. Then O' is the reflection of O in the plane ABC, and OO' cuts this plane at the centroid G of ΔABC. Hence

$$\vec{OO'} = 2\vec{OG} = \frac{2}{3}(\bar{i} + \bar{j} + \bar{k}).$$

On the other hand

$$\vec{O'O} = \bar{i}' - \bar{i} = \bar{j}' - \bar{j} = \bar{k}' - \bar{k},$$

whence

$$\bar{i}' = \bar{i} - \vec{OO'} = \frac{1}{3}\bar{i} - \frac{2}{3}\bar{j} - \frac{2}{3}\bar{k},$$

and something similar for $\bar{j}', \bar{k}'$.

Therefore, we have the relations between the two bases, and the old coordinates of O'. If we proceed as for Proposition 1.3.12, we obtain the answer.

1.4.1. In ΔABC, let $BB' \perp AC$, $CC' \perp AB$, and $H = BB' \cap CC'$. If H is chosen as the origin, the previous hypotheses mean

$$\bar{r}_C \cdot (\bar{r}_B - \bar{r}_A) = 0, \quad \bar{r}_B \cdot (\bar{r}_A - \bar{r}_C) = 0.$$

Adding these relations we get $\bar{r}_A.(\bar{r}_B - \bar{r}_C) = 0$, i.e., AH is the third height.

1.4.2. If the sides are $\bar{a} = \bar{a}'$, $\bar{b} = \bar{b}'$, the diagonals are $\bar{a} + \bar{b}$, $\bar{a} - \bar{b}$, and

$$(\bar{a} + \bar{b})^2 + (\bar{a} - \bar{b})^2 = 2\bar{a}^2 + 2\bar{b}^2.$$

1.4.3. $\vec{PQ} = \vec{OQ} - \vec{OP}$, and if $\bar{u} \in d$ then $\vec{PQ} \cdot \bar{u} = \vec{OQ} \cdot \bar{u} - \vec{OP} \cdot \bar{u} = 0$, in view of the hypotheses.

1.4.4. Denote by A, B, C, D the vertices and choose A as origin. Then if, say

$$\bar{r}_B \cdot (\bar{r}_D - \bar{r}_C) = 0, \quad \bar{r}_C \cdot (\bar{r}_B - \bar{r}_D) = 0,$$

it follows by adding these two relations that $\bar{r}_D \cdot (\bar{r}_B - \bar{r}_C) = 0$.

1.4.5. We have $\vec{PQ}(6,7)$, $\vec{QR}(6,-2)$, $\vec{RP}(-12,-5)$, and their lengths and angles will be computed by formulas (1.4.8), (1.4.9).

1.4.6. Use formulas (1.4.8), (1.4.9) for the vectors $\bar{v}$ and $\vec{PQ}(3,6,2)$. The required projection is

$$pr_{\vec{PQ}}\bar{v} = (mpr_{\vec{PQ}}\bar{v})\frac{\vec{PQ}}{|\vec{PQ}|} = \left(\frac{\bar{v} \cdot \vec{PQ}}{|\vec{PQ}|}\right)\frac{\vec{PQ}}{|\vec{PQ}|}.$$

1.4.7. Assume that the first frame has $O = A$, $\bar{i} = (1/2)\vec{AB}$, $\bar{j} = (1/2)\vec{AD}$, and the second has $O' = AC \cap BD$, $\bar{i}' = (1/\sqrt{2})\vec{O'B}$, $\bar{j}' = (1/\sqrt{2})\vec{O'C}$. Then

$$\vec{OO'} = \bar{i} + \bar{j}, \; \bar{i}' = \frac{1}{\sqrt{2}}(\bar{i} - \bar{j}), \; \bar{j}' = \frac{1}{\sqrt{2}}(\bar{i} + \bar{j}),$$

and we will proceed in accordance with Section 1.3.

1.4.8. Take O as the origin. Then the vectors

$$\bar{v}_1 = \bar{r}_A \times (\bar{r}_B \times \bar{r}_C), \; \bar{v}_2 = \bar{r}_B \times (\bar{r}_C \times \bar{r}_A), \; \bar{v}_3 = \bar{r}_C \times (\bar{r}_A \times \bar{r}_B)$$

are perpendicular to π_A, π_B, π_C, respectively. Using formula (1.4.35) we get $\bar{v}_1 + \bar{v}_2 + \bar{v}_3 = \bar{0}$. Hence $\bar{v}_1 \times \bar{v}_2 \perp \bar{v}_1, \bar{v}_2, \bar{v}_3$, and the line through O in the direction of $\bar{v}_1 \times \bar{v}_2$ belongs to π_A, π_B, π_C.

1.4.9. We have

$$\bar{n}_A = \frac{1}{2}\vec{CD}\times\vec{CB}, \bar{n}_B = \frac{1}{2}\vec{CA}\times\vec{CD}, \bar{n}_C = \frac{1}{2}\vec{BD}\times\vec{BA}, \bar{n}_D = \frac{1}{2}\vec{BA}\times\vec{BC}.$$

Then, the result is checked easily by using A as origin.

1.4.10. By formula (1.4.26) we get $\bar{a} \times \bar{b} = -2\bar{i} - 3\bar{j} - 7\bar{k}$, and the requested area is $|\bar{a} \times \bar{b}| = \sqrt{62}$.

1.4.11. We get

$$(\bar{a}, \bar{b}, \bar{c}) = -49, \ \bar{a} \times \bar{b} = 3\bar{i} - 17\bar{j} - 5\bar{k}.$$

Hence, the height is $49/\sqrt{323}$.

1.4.12. The equations yield

$$\bar{x} \cdot (\beta\bar{a} - \alpha\bar{b}) = 0, \ \bar{x} \cdot (\gamma\bar{a} - \alpha\bar{c}) = 0,$$

whence

$$\bar{x} = \lambda(\beta\bar{a} - \alpha\bar{b}) \times (\gamma\bar{a} - \alpha\bar{c}).$$

Now, λ is determined from $\bar{x} \cdot \bar{a} = \alpha$, and one gets

$$\bar{x} = \frac{\alpha(\bar{b} \times \bar{c}) + \beta(\bar{c} \times \bar{a}) + \gamma(\bar{a} \times \bar{b})}{(\bar{a}, \bar{b}, \bar{c})}.$$

2.2.1. The coordinates of $P = d_1 \cap d_2$ are the solutions of the given system of equations: $P(3,1)$. The coordinates of M are given by Example 1.3.10 i.e., $M(2,1)$. Hence, the required line PM has the equation $y = 1$.

2.2.2. CD and CB are symmetric with AB, AD, respectively, with respect to the origin. Hence, their equations will be obtained by changing the sign of x, y in the given equations, i.e.,

$$(CD) \quad -x - 2y - 1 = 0, \ (CA) \quad -x - y + 1 = 0.$$

Now, it is easy to find the coordinates of A, B, C, D, and the equations of the diagonals.

The general case can be treated similarly, after first performing the coordinate transformation $\tilde{x} = x - x_0$, $\tilde{y} = y - y_0$.

Another way to proceed is by denoting $A(\lambda, \mu)$, where λ, μ are the solutions of the given equations of (AB), (AD). Then, since $(A, M; C) = -2$, C has the coordinates $\lambda' = 2x_0 - \lambda$, $\mu' = 2y_0 - \mu$. Since $CD \| AB$, CD has an equation of the form $ax + by + c' = 0$ which is satisfied by the coordinates of C. This yields $c' = -c - 2(ax_0 + by_0)$. In a similar way, we get the equation of CD, etc.

2.2.3. i) If M is the midpoint of BC, $M(0, 1, 3/2)$, and

$$(AM) \qquad \frac{x-1}{-1} = \frac{y}{1/2} = \frac{z}{1/2}.$$

The height is the intersection of the planes

$$\text{plane}(ABC): \; x + \frac{1}{2}y + \frac{1}{3}z - 1 = 0, \quad \text{plane}(A \perp BC): \; 2y - 3z = 0.$$

For the bisector of $\hat{A}$, let N, M be points of AB, AC such that

$$\vec{AN} = \frac{\vec{AB}}{|\vec{AB}|}, \qquad \vec{AM} = \frac{\vec{AC}}{|\vec{AC}|}.$$

Hence, $\vec{AN}(-1/\sqrt{5}, 2/\sqrt{5}, 0)$, $\vec{AM}(-1/\sqrt{10}, 0, 3/\sqrt{10})$, and $\vec{NM}(3/\sqrt{10}, -2/\sqrt{5}, 3/\sqrt{10})$. The bisector is the intersection of the plane ABC (equation above) with

$$\text{plane}(A \perp \vec{NM}): \quad \frac{3}{\sqrt{10}}(x-1) - \frac{2}{\sqrt{5}}y + \frac{3}{\sqrt{10}}z = 0.$$

The perpendicular bisector of BC is the intersection of

$$\text{plane}(ABC)\, x + \frac{1}{2}y + \frac{1}{3}z - 1 = 0; \; \text{plane}(M \perp BC)\, 2(y-1) + 3(z - \frac{3}{2}) = 0.$$

ii) The centroid G is $G(1/3, 2/3, 1)$, and the other required points are obtained by writing the equations of two heights, bisectors, etc. and finding the coordinates of their intersection points.

iii) Use the same method as in i).

2.2.4. Notice that $A_1(-3,5,0) \in d_1$ and $A_2(10,-7,0) \in d_2$. d_1 has the direction vector $\bar{v}_1(2,3,1)$ and d_2 has the direction vector $\bar{v}_2(5,4,1)$. Finally, the third line has direction vector $\bar{w}(8,7,1)$. The required line is the intersection of

$$\text{plane}(A_1,\bar{v}_1,\bar{w})\, 2x-3y+5z+21=0;\ \text{plane}(A_2,\bar{v}_2,\bar{w})\, x-y-z-17=0.$$

The intersection points are the solutions of the corresponding equations. The direction of d is that of the vector product of the normals of the two planes written above, i.e., of $(2,-3,5)$ by $(1,-1,-1)$, and the result is $8\bar{i}+7\bar{j}+\bar{k}$. Hence, the direction parameters are $(8,7,1)$, and the cosines are $(8/\sqrt{114}, 7/\sqrt{114}, 1/\sqrt{114})$.

2.2.5. Let $M(x_0,y_0,z_0)$ be the midpoint of PP'. Then $M \in d$, $PM \perp d$, i.e.,

$$\frac{x_0-1}{2}=\frac{y_0-2}{4}=\frac{z_0-3}{5},$$

$$2(x_0-4)+3(y_0-3)+5(z_0-10)=0.$$

This yields $x_0=129/41$, $y_0=258/41$, $z_0=343/41$. Now, the coordinates of P' are given by $(P,M;P')=-2$, and we have $P'(-94/41, -393/41, -276/41)$.

2.2.6. We have $\vec{AB}(-1,2,-2)$, $\vec{BC}(-1,2,2)$, $\vec{CD}(1,-2,2)$, $\vec{DA}(1,-2,-2)$. Hence $ABCD$ has parallel opposite sides, and all of equal length. It is easy to write the required equations. In particular

$$(AC)\ \ 2x+y=1,\ \ z=0;\qquad (BD)\ \ x=0,\ \ y=1.$$

Hence, the new origin is $O'=AC\cap BD$ of coordinates $O'(0,1,0)$. The new basis is

$$\bar{i}'=\frac{\vec{AC}}{|\vec{AC}|},\qquad \bar{j}'=\frac{\vec{BD}}{|\vec{BD}|},\qquad \bar{k}'=\bar{i}'\times\bar{j}'.$$

These vectors are easily computed, and the required coordinate transformation will be written as in Section 1.3.

2.2.7. Use an affine frame with the origin at A, and basic vectors equal to those of the edges which begin at A. Then the diagonal is $x = y = z$, and the plane is $x + y + z = 1$. The intersection point has coordinates $(1/3, 1/3, 1/3)$.

2.2.8. If $A(a,0,0)$, $B(0,b,0)$, $C(0,0,c)$, the centroid is $G(a/3, b/3, c/3)$. Hence, if A is fixed the locus is the plane $x = a/3$, and if B, C are fixed the locus is the line $y = b/3$, $z = c/3$.

2.2.9. If the required projection is $\bar{v}(v_1, v_2, v_3)$, $\bar{v} - \bar{u}$ is normal to the plane and this yields

$$v_1 = \alpha + a\lambda, \quad v_2 = \beta + b\lambda, \quad v_3 = \gamma + c\lambda \quad (\lambda \in \mathbf{R}).$$

Then, λ is determined by the condition that $\bar{v}$ belongs to the plane $av_1 + bv_2 + cv_3 = 0$.

2.3.1. We have $A(1,0)$, $B(0,-2)$, and there are two possible positions C_1, C_2 of C, which are symmetric with respect to AB. (Draw a figure.) Generally, formula (2.3.10) yields the counterclockwise angle $\varphi \in [0, \pi]$ from d_1 to d_2. Since the slope $m_{AB} = 2$, we have

$$\frac{1}{2} = \frac{m_{AC_1} - 2}{1 + 2m_{AC_1}}, \quad \frac{1}{2} = \frac{2 - m_{AC_2}}{1 + 2m_{AC_2}},$$

$$\frac{4}{3} = \frac{2 - m_{BC_1}}{1 + 2m_{BC_1}}, \quad \frac{4}{3} = \frac{m_{BC_2} - 2}{1 + 2m_{BC_2}}.$$

Therefore, $m_{AC_1} = \infty$, $m_{AC_2} = 3/4$, $m_{BC_1} = 2/11$, $m_{BC_2} = -2$, and (AC_1) $x = 1$, (BC_1) $y + 2 = (2/11)x$, (AC_2) $y = (3/4)(x - 1)$, (BC_2) $y + 2 = -2x$.

2.3.2. We have $4/2 = (-6)/(-3) \neq (-3)/7$. Hence, the lines are parallel. They cut the y-axis at $A(0, -1/2)$, $B(0, 7/3)$, respectively, and the midpoint of AB is $M(0, 11/12)$. The required line is

$$y - \frac{11}{12} = \frac{2}{3}x.$$

2.3.3. If we choose the line BC as the x-axis, and the origin at the midpoint of BC, we have $B(b,0)$, $C(-b,0)$ ($b = const.$), and the required locus of $A(x,y)$ is

$$(x - b)^2 + y^2 - (x + b)^2 - y^2 = k \, (= const.)$$

i.e., $x = -k/4b$. This is a straight line perpendicular to BC.

2.3.4. Take the triangle as ΔABC, and BC as the fixed line. Take BC as the x-axis and $AO \perp BC$ as the y-axis. Denote $A(0,a)$, $B(b,0)$, $C(0,c)$. Take $P(p,0)$ as the first vertex of the rectangle $PNMQ$, where $P, Q \in BC$, $N \in CA$, $M \in AB$. Then

$$(AC)\ \ \frac{x}{c} + \frac{y}{a} - 1 = 0, \quad (AB)\ \ \frac{x}{b} + \frac{y}{a} - 1 = 0,$$

and $N(p, a(c-p)/c)$, $M(bp/c, a(c-p)/c)$, $Q(bp/c, 0)$. The center of the rectangle is the midpoint of PM and it has the coordinates

$$x = \frac{p(c+b)}{2c}, \quad y = \frac{a(c-p)}{2c}.$$

If $c + b = 0$ (i.e., ΔABC is an isosceles triangle), the locus is $x = 0$. Otherwise the equation of the locus is given by solving p from x and inserting the result into y. The result is the line

$$y = -\frac{a}{b+c}x + \frac{a}{2}.$$

2.3.5. Choose affine coordinates with the origin at O, the x-axis along OA, and the y-axis along OC. Then the coordinates of the various points may be denoted as follows:

$$A(\alpha, 0),\ B(\beta, m\beta),\ C(0,\gamma),\ P(p_1,p_2),\ Q(q_1,q_2),$$

where α, β, γ are variable, and m, p_1, p_2, q_1, q_2 are constant. We have:

$$(PB)\ \ \frac{x-p_1}{\beta - p_1} = \frac{y-p_2}{m\beta - p_2}, \quad (QB)\ \ \frac{x-q_1}{\beta-q_1} = \frac{y-q_2}{m\beta - q_2}.$$

Since $A \in PB$ and $C \in QB$, we get

$$\alpha = \frac{\beta(mp_1 - p_2)}{m\beta - p_2}, \quad \gamma = \frac{\beta(q_2 - mq_1)}{\beta - q_1},$$

whence

$$(AC)\ \ \beta\left[\frac{x}{mp_1 - p_2} + \frac{y}{q_2 - mq_1} - 1\right] - \left[\frac{p_2 x}{mp_1 - p_2} + \frac{q_1 y}{q_2 - mq_1}\right] = 0.$$

Since this is the equation of a pencil of lines, we are done.

The reader is advised to give a more straightforward solution of this problem by using Desargues' theorem of Section 5.1.

2.3.6. We may represent $\bar{r}$ by (2.2.18), and the values of t and s are internal coordinates in the plane of $\Delta M_1M_2M_3$ for the frame with origin M_1 and basic vectors $\overrightarrow{M_1M_2}$, $\overrightarrow{M_1M_3}$. These coordinates have a known interpretation as simple ratios (Proposition 1.3.7), which implies that P is interior to the triangle iff $0 < s < 1$, $0 < t < 1$. But then, $0 < 1 - s - t < 1$, and (2.2.18) becomes the required result.

2.3.7. If we use an affine frame where $A(0,0)$, $B(1,0)$, $C(0,1)$, we have $M_3(\lambda, 0)$, $M_2(0, \mu)$, $M_1(\alpha, 1-\alpha)$. Accordingly: $\rho_1 := (B, C; M_1) = (1-\alpha)/\alpha$, $\rho_2 := (C, A; M_2) = (1-\mu)/\mu$, $\rho_3 := (A, B; M_3) = \lambda/(1-\lambda)$, and

$$\alpha = \frac{1}{1+\rho_1}, \quad \mu = \frac{1}{1+\rho_2}, \quad \lambda = \frac{\rho_3}{1+\rho_3}.$$

Now, a computation shows that M_1, M_2, M_3 are collinear iff $\rho_1\rho_2\rho_3 = -1$, and AM_1, BM_2, CM_3 are concurrent iff $\rho_1\rho_2\rho_3 = 1$.

2.3.8. If φ is the angle between the planes $x-z+4=0$, $x-4y-8z+12=0$, we get $\cos\varphi = 1/\sqrt{2}$, and $\varphi \neq \pi/6$. Thus, the required plane has an equation of the form

$$x + 5y + z + \lambda(x - z + 4) = 0,$$

where λ is determined by the condition that the angle of the vectors $(1+\lambda, 5, 1-\lambda)$, $(1, -4, -8)$ is either $\pi/6$ or $5\pi/6$. This yields

$$\pm\frac{\sqrt{3}}{2} = \frac{-3+\lambda}{\sqrt{27+2\lambda^2}},$$

whence $\lambda = (-12 \pm \sqrt{54})/2$.

2.3.9. Use formula (2.3.13).

2.3.10. Use the same method as for Problem 2.3.8.

2.3.11. The projection of d onto the (x, y)-plane is the intersection of the latter with the plane which contains d and is parallel to Oz. The last plane is

$$3x + 2y - z + 5 + \lambda(x - y - z + 1) = 0,$$

where $-1-\lambda=0$. Thus, the projection is $2x+3y+4=0$, $z=0$. The two other required projections will be obtained similarly.

2.3.12. We have

$$(AB)\quad \frac{x-1}{-1}=\frac{y-1}{0}=\frac{z}{1}$$

i.e., $y-1=0$, $x+z-1=0$. Thus the required plane is

$$x+z-1+\lambda(y-1)=0,$$

where $(1,\lambda,1)\perp \vec{CD}=(-2,1,0)$ i.e., $\lambda=2$.

2.3.13. Use (2.3.14), i.e.:

$$\delta=\frac{|\vec{BC}\times\vec{BA}|}{|\vec{BC}|}.$$

2.3.14. The required line is the intersection of the plane α_1 which contains the point A, a point of the fixed line e.g., $B_1(1,-3,5)$, and the direction vector of the first line $\bar{v}_1(2,4,3)$, with the plane α_2 through A, $B_2(0,2,-1)$, $\bar{v}_2(5,-1,2)$. The equations of these planes are

$$(\alpha_1)\ \ 25x-17y+6z-106=0,\qquad (\alpha_2)\ \ x-y-3z-1=0.$$

2.3.15. The required line is the intersection of the plane α_1 which contains A and the given line, and the plane α_2 which contains A and is orthogonal to $\bar{v}(2,-1,3)$. And we have

$$(\alpha_2)\ \ 2(x-2)-(y-3)+3(z-1)=0,$$

and α_1 is the plane of the pencil

$$x+2y+1+\lambda(3y+z-2)=0$$

through the given line, which is satisfied by $(2,3,1)$. This yields $\lambda=-9/8$ and

$$(\alpha_1)\ \ 8x-11y-9z+26=0.$$

2.3.16. The equations of d_1 and d_2 have the common solution $x=1$, $y=2$, $z=3$, and this defines a common point A of the two lines. The plane which contains d_1 and d_2 is determined by the point A and the vectors $\bar{v}_1(2,-2,1)$, $\bar{v}_2(11,10,2)$, and its equation is

$$2x - y - 3z + 9 = 0.$$

For the second question, we compute the angle of the direction vectors $\bar{v}_1$ and $\pm\bar{v}_2$. We get $\cos(\bar{v}_1, -\bar{v}_2) < 0$, i.e., the obtuse angle exists and it is between $\bar{v}_1$ and $-\bar{v}_2$. In order to obtain the required bisector we take the points $B_1 \in d_1$, $B_2 \in d_2$ of radius vectors

$$\bar{r}_{B_1} = \bar{r}_A + \frac{\bar{v}_1}{|\bar{v}_1|}, \bar{r}_{B_2} = \bar{r}_A - \frac{\bar{v}_2}{|\bar{v}_2|},$$

and the midpoint M of B_1B_2. The bisector is the line AM. The results are $B_1(5/3, 4/3, 10/3)$, $B_2(4/15, 20/15, 43/15)$, $M(29/30, 40/30, 83/30)$, with corresponding equations for AM.

2.3.17. Use (2.3.15) and the method described in the proof of Proposition 2.3.3. In our case we have $\bar{r}_1(-3,5,0)$, $\bar{r}_2(-2,1,3)$, $\bar{v}_1(2,3,1)$, $\bar{v}_2(8,7,1)$.

2.3.18. i) Take the origin at O and the axes through A, B, C, respectively, such that $A(a,0,0)$, $B(0,b,0)$, $C(0,0,c)$. Then the required planes have the equations

$$by - cz = 0, \quad cz - ax = 0, \quad ax - by = 0,$$

and since

$$ax - by = -(by - cz) - (cz - ax),$$

the planes belong to a pencil.

ii) It is clear that the common line of the three planes of i) is the perpendicular from O to plane(ABC), while the intersection of the former planes with plane(ABC) are the heights of ΔABC.

3.1.1. i) The center of the circle is $A(1,0)$, the radius is $\rho = 2$, and the distance $AP = \sqrt{5}/2$. d is the perpendicular from P to AP hence, d has the equation $4x - 2y - 9 = 0$.

ii) For $M(x_0, y_0)$ the polar line is

$$x(x_0 - 1) + yy_0 - (x_0 + 3) = 0.$$

Since this is d, we must have

$$\frac{x_0 - 1}{4} = \frac{y_0}{-2} = \frac{x_0 + 3}{9},$$

whence $x_0 = 21/5$, $y_0 = -8/5$.

The tangents have the equation (3.1.19) i.e.,

$$[x(x_0 - 1) + yy_0 - (x_0 + 3)]^2 - [(x-1)^2 + y^2 - 4][(x_0 - 1)^2 + y_0^2 - 4] = 0,$$

etc.

3.1.2. Use (3.1.19).

3.1.3. The radical axis of the circles is $x = 0$, and it cuts the circles at $P(0,4)$, $Q(0,-4)$. Hence, we are asked to compute the angle of the vectors $\vec{OP}$, $\vec{CP}$, where $C(5,0)$ is the center of the second circle.

3.1.4. If the circles are

$$(\Gamma_1)\ \ (\bar{r} - \bar{a}_1)^2 - \rho_1^2 = 0, \quad (\Gamma_2)\ \ (\bar{r} - \bar{a}_2)^2 - \rho_2^2 = 0,$$

where $\bar{a}_1$, $\bar{a}_2$ are the radius vectors of the centers, the radical axis is

$$(d)\ \ 2\bar{r} \cdot (\bar{a}_2 - \bar{a}_1) + \bar{a}_1^2 - \bar{a}_2^2 = 0,$$

and it is orthogonal to $\bar{a}_2 - \bar{a}_1$. Furthermore, since $M \in d$ has equal power l^2 with respect to Γ_1 and Γ_2, the tangents from M to Γ_1 and Γ_2 are of equal length l, and the circle of center M and radius l is orthogonal to Γ_1 and Γ_2.

3.1.5. Use Problem 3.1.4. The radical axis of the circles is $3x + y = 0$ hence, the center of Γ is of the form $C(-3\alpha, \alpha)$ and

$$(\Gamma)\ \ (x + 3\alpha)^2 + (y - \alpha)^2 = 9\alpha^2 + (\alpha + 1)^2 - 16.$$

Since this equation is satisfied by $M(5,-4)$, we get $\alpha = -14/9$, and

$$(\Gamma)\ \ (x + \frac{14}{3})^2 + (y + \frac{14}{9})^2 = 77/9.$$

3.1.6. Choose AB, CD as x-axis, y-axis, respectively, and $C(0,a)$, $D(0,-a)$. Then $(\Gamma)\ \ x^2 + y^2 = a^2$, $(d)\ \ y = mx + a$, where m is a parameter. This yields

$$N(-\frac{2am}{1+m^2}, \frac{a(1-m^2)}{1+m^2}),$$

with the tangent line

$$2mx + (m^2 - 1)y + a(1 + m^2) = 0.$$

For the intersection of this tangent with $x = -a/m$ we get $y = -a$, as required.

3.1.7. Take BC as Ox such that $B(b, 0)$, $C(-b, 0)$. Denote $A(\alpha, \beta)$. Then the centroid is $G(x = \alpha/3,\ y = \beta/3)$. The slopes of AB, AC are

$$m_{AB} = \frac{\beta}{\alpha - \beta}, \qquad m_{AC} = \frac{\beta}{\alpha + \beta},$$

and, if we use formula (2.3.10), the constancy of $\widehat{BAC}$ means

$$\frac{\frac{\beta}{\alpha+\beta} - \frac{\beta}{\alpha-\beta}}{1 + \frac{\beta^2}{\alpha^2-\beta^2}} = \pm k,$$

where $k = const. > 0$. By inserting here $\alpha = 3x$, $\beta = 3y$, we see that the locus consists of two arcs of circles, above and under the side BC.

3.1.8. Take the origin at the center of Γ, and the equation of Γ : $x^2 + y^2 - \rho^2 = 0$. If $M(x_0, y_0)$, the equation of the tangents of Γ through M is (3.1.19) i.e.,

$$(xx_0 + yy_0 - \rho^2)^2 - (x^2 + y^2 - \rho^2)(x_0^2 + y_0^2 - \rho^2) = 0.$$

After computations, this equation becomes

$$(\rho^2 - y_0^2)(x - x_0)^2 + (\rho^2 - x_0^2)(y - y_0)^2 + 2x_0y_0(x - x_0)(y - y_0) = 0.$$

The slope of a tangent is $m = (y - y_0)/(x - x_0)$. Hence, for the two tangents we have

$$m_1m_2 = \frac{\rho^2 - y_0^2}{\rho^2 - x_0^2} = -1.$$

Thus, the locus is the circle $x_0^2 + y_0^2 = 2\rho^2$.

3.1.9. Choose the origin at P, and Ox along PC, where C is the center of Γ. Then

$$(\Gamma)\ \ x^2 + y^2 - 2\rho x = 0,$$

and it cuts (d) $y = mx$ (m is a parameter) at $P(0,0)$ and $Q(2\rho/(1+m^2), 2m\rho/(1+m^2))$. The condition $(P, M; Q) = \lambda$ is equivalent to

$$x_M = cx_Q, \quad y_M = cy_Q,$$

where $c = (1+\lambda)/\lambda$. Hence, the locus of M is the result of the elimination of m from

$$x = \frac{2c\rho}{1+m^2}, \quad y = \frac{2mc\rho}{1+m^2}$$

i.e.,

$$x^2 + y^2 - 2c\rho = 0.$$

3.1.10. The equation of the sphere must be of the form

$$x^2 + y^2 + (z-\lambda)^2 = \rho^2,$$

and the conditions stated mean

$$\rho^2 - \lambda^2 = 16, \quad 9 + (1-\lambda)^2 = \rho^2.$$

Therefore: $\lambda = -3$, $\rho = 5$.

3.1.11. If the pole is $P(x_0, y_0, z_0)$, the polar plane is $xx_0 + yy_0 + zz_0 - \rho^2 = 0$, and

$$\frac{x_0}{a} = \frac{y_0}{b} = \frac{z_0}{c} = -\frac{\rho^2}{d},$$

defines x_0, y_0, z_0. The plane is tangent if the obtained values satisfy $x_0^2 + y_0^2 + z_0^2 = \rho^2$.

3.1.12. i) The centers and radii are $[(0,0,0),3]$ and $[(0,-2,3),\sqrt{6}]$, respectively. ii) We want λ such that

$$x^2 + y^2 + z^2 - 9 + \lambda(x^2 + y^2 + z^2 + 4y - 6z + 7) = 0$$

is satisfied by $(1,1,-1)$. This yields $\lambda = 3/10$, and the sphere is

$$x^2 + y^2 + z^2 + (12/13)y - (18/13)z - 69/13 = 0.$$

3.1.13. Look for λ such that the distance from the center $(-5,8,-1)$ to the plane $y + \lambda z = 0$ equals the radius 4. This means

$$\frac{|8-\lambda|}{\sqrt{1+\lambda^2}} = 4,$$

and $\lambda_1 = 12/5$, $\lambda_2 = -4/3$.

3.1.14. The center of the sphere is the common point of three planes: the planes which are orthogonal to the given lines at $M(1, -4, 6)$, $N(4, -3, 2)$, respectively, and the orthogonal bisector plane of the segment MN. The equations of these planes are

$$3(x-1)+6(y+4)+4(z-6)=0,$$

$$2(x-4)+(y+3)-6(z-2)=0,$$

$$3(x-\frac{5}{2})+(y+\frac{7}{2})-4(z-4)=0,$$

which yields the center $C(-5, 3, 0)$. The radius is $CM = 11$. Thus the result is

$$(x+5)^2+(y-3)^2+z^2=121.$$

3.1.15. If $P(x, y, z)$, it must satisfy

$$(x-5)^2+(y-3)^2+(z+1)^2-9=7,$$

$$(x-7)^2+y^2+(z-2)^2-16=7.$$

Hence, being the intersection of two spheres, the locus is a circle.

3.1.16. Take the frame such that d is the z-axis, and the center C of Γ is on the x-axis i.e., $C(a, 0, 0)$. Then the planes α through d have the equation $x + \lambda y = 0$, and the intersection circle has the center at the intersection of α with the perpendicular line from C to α i.e.,

$$\frac{x-a}{1}=\frac{y}{\lambda}=\frac{z}{0}.$$

The elimination of λ yields

$$x^2+y^2-ax=0, \quad z=0,$$

which is a circle. Of course, the true locus is only the part of this circle which is inside Γ.

If we have the point A, rather than a line d, we may choose the frame such that $A(0, 0, 0)$, $C(a, 0, 0)$, and the plane α has the equation

$$x+\lambda y+\mu z=0.$$

As above,we get the equation of the locus under the form

$$x^2+y^2+z^2-ax=0.$$

3.2.1. If divided by 4225, the equation becomes

$$\frac{x^2}{13^2}+\frac{y^2}{5^2}=1.$$

Thus $a=13$, $b=5$, $c=12$, $e=12/13$, $d=169/12$.

3.2.2. Since $AP=AQ$, P, Q are at the intersection of the ellipse with a circle of center A. Since the x-axis is a symmetry axis for both curves, P, Q must be symmetric points, and their coordinates are of the form

$$P(\lambda,\frac{1}{2}\sqrt{36-\lambda^2}),\quad Q(\lambda,-\frac{1}{2}\sqrt{36-\lambda^2}).$$

Now, we still have to ask $PQ=AP$, which amounts to

$$36-\lambda^2=(6-\lambda)^2+\frac{1}{4}(36-\lambda^2).$$

This simplifies by $6-\lambda$, and the remaining equation yields $\lambda=6/7$. Therefore, $P(6/7,12\sqrt{3}/7)$, $Q(6/7,-12\sqrt{3}/7)$.

3.2.3. Take the axes such that the hyperbola is $xy=k$, and denote $A(a_1,a_2)$, $B(b_1,b_2)$, $C(c_1,c_2)$ $(a_1a_2=b_1b_2=c_1c_2=k)$. The heights of ΔABC which start at A, B have the equations

$$(c_1-b_1)(x-a_1)+(c_2-b_2)(y-a_2)=0,$$

$$(c_1-a_1)(x-b_1)+(c_2-a_2)(y-b_2)=0.$$

The solutions of this system of equations are

$$x=-\frac{k^2}{a_1b_1c_1},\quad y=-\frac{a_1b_1c_1}{k},$$

and they satisfy the equation $xy=k$.

3.2.4. We have $a=7$, $b=\sqrt{24}$, hence, $c=5$, and the "basis" of Δ equals 10. Then, the side of Δ of equation $x=5$ cuts the ellipse at

$y = \pm 24/7$. Thus, the height of Δ equals 48/7, and the area of Δ is 480/7.

3.2.5. Denote the coordinates $A(a, 0)$, $B(0, b)$. Then the coordinates of M are

$$x = \frac{a}{1+k}, \quad y = \frac{kb}{1+k},$$

and the condition $a^2 + b^2 = l^2$ yields the equation of the locus of M

$$\frac{x^2}{l^2/(1+k^2)} + \frac{y^2}{k^2l^2/(1+k^2)} = 1.$$

The locus is an ellipse.

3.2.6. With the notation of Section 3.2, we have $a = 3$, $b = 4$, $c = \sqrt{a^2+b^2} = 5$, $e = c/a = 5/3$, $d = a/e = 9/5$.

3.2.7. The hyperbola is $x^2/a^2 - y^2/b^2 = 1$, where $b/a = 1/2$, and it is satisfied by $(12, \sqrt{3})$. This yields $b^2 = 33$, $a^2 = 4b^2 = 132$.

3.2.8. With the same notation (e.g., Exercise 3.2.7), P is at the intersection of the lines

$$y = \frac{b}{a}x, \qquad y = -\frac{a}{b}(x-c).$$

This system yields

$$x = \frac{a^2c}{a^2+b^2} = \frac{a^2}{c} = d.$$

3.2.9. If $P(x', y')$, $M(x = x', y = \lambda y')$, and we have $x^2 + y^2/\lambda^2 = 1$. Hence the locus is an ellipse.

3.2.10. If the given circles have the centers C_1, C_2 and the radii ρ_1, ρ_2, and if the common tangent circles have center C and radius ρ, we must have one of the relations

$$|CC_1 \mp CC_2| = |(\rho \pm \rho_1) \pm (\rho \pm \rho_2)| = |\rho_1 \pm \rho_2|.$$

Hence the locus consists of hyperbolas or ellipses in agreement with the relative position of the given circles.

3.2.11. If P is the intersection of $x = 16/5$, $y = (b/a)x$, we have

$$d = \frac{a^2}{\sqrt{a^2+b^2}} = 16/5, \quad \frac{b}{a} = \frac{12}{16}.$$

This yields $a = 4$, $b = 3$.

3.2.12. Because of symmetry, if (α, β) is a vertex of the square, the other vertices must be $(-\alpha, \beta)$, $(-\alpha, -\beta)$, $(\alpha, -\beta)$, and $\alpha^2/a^2 - \beta^2/b^2 = 1$, $\alpha = \beta$. The solution is $\alpha = ab/\sqrt{b^2 - a^2}$, and it exists iff $b > a$.

3.2.13. We have $A(0,0)$, and the orthocenter $M = F(4,0)$. Suppose $B(\lambda^2/8, \lambda)$, $C(\mu^2/8, \mu)$. If we write that $CF \perp AB$, $BF \perp AC$, we get

$$\mu^2 - 32 = -\lambda\mu, \quad \lambda^2 - 32 = -\lambda\mu.$$

Hence, $\lambda = \mu = 4$. Now, we know the coordinates of the vertices of ΔABC, and it is easy to write down the equations of its sides.

3.2.14. Let $M_i(x_i, y_i)$ $(i = 1,2,3)$ $(x_i = y_i^2/2a)$ be the vertices of the triangle. It is not possible to have $y_i > 0$ for all $i = 1,2,3$ since then, if also $y_1 < y_2 < y_3$ (for instance), we would have that $M_1\widehat{M_2}M_3$ is obtuse. It is also impossible to have, say, $y_1 < 0$, y_2, $y_3 > 0$, since the distance from M_1 to the points of the upper part of the parabola increase as the upper point runs from O to ∞ along the parabola, and we cannot have $M_1M_2 = M_1M_3$. Thus, say, $M_1 = O$ i.e., $y_1 = 0$, and then M_2, M_3 must be symmetric with respect to the x-axis i.e., $-y_3 = y_2 > 0$. And, $\Delta M_1M_2M_3$ is equilateral iff

$$2y_2 = OM_2 = \sqrt{\frac{y_2^4}{4a^2} + y_2^2},$$

which gives $y_2 = 2a\sqrt{3}$. The length of the side of $\Delta M_1M_2M_3$ is $4a\sqrt{3}$.

3.2.15. The (y, z)-equation of the intersection ellipse is

$$\frac{y^2}{b^2a^2/(a^2 - k^2)} + \frac{z^2}{c^2a^2/(a^2 - k^2)} = 1.$$

Hence, the eccentricity is $e = \sqrt{c^2 - b^2}/c$, if $c \geq b$, and $e = \sqrt{b^2 - c^2}/b$, if $c \leq b$, and $e = const.$

3.2.16. The elimination of z between the two given equations yields the equation of a cylinder with generators parallel to Oz, and which contains the intersection of the plane with the hyperboloid. Namely, we get

$$\frac{xy}{3} - \frac{2x}{3} - y + 2 = 0.$$

Hence, this is the (x, y)-equation of the projection of the intersection line on the plane $z = 0$. But this equation may also be written as

$$(\frac{x}{3} - 1)(y - 2) = 0$$

hence it represents a pair of straight lines. Accordingly, the intersection of the plane and the hyperboloid must also consist of two straight lines.

3.2.17. The lines d which generate the required locus may be seen as defined by an arbitrary point of the first line, $M(2t, 1, -t)$ (t is a parameter), and which meet the second and third lines. Accordingly, we get the following pair of equations of d:

$$\begin{vmatrix} x-2t & y-1 & z+t \\ 2t-2 & 1 & -t \\ 0 & 1 & 1 \end{vmatrix} = 0, \quad \begin{vmatrix} x-2t & y-1 & z+t \\ 2t & 2 & -t \\ 2 & 0 & 1 \end{vmatrix} = 0.$$

The computation of the determinants yields

$$x + 2y - 2z - 2 + t(x - 2y + 2z - 2) = 0,$$

$$x - 2z - 2t(y + 1) = 0,$$

and then, the elimination of t provides the equation of the locus

$$\frac{x^2}{4} + \frac{y^2}{1} - \frac{z^2}{1} - 1 = 0.$$

The locus is a one-sheeted hyperboloid.

3.2.18. If $y = 2$ is inserted in the equation of the paraboloid, one gets $x^2 = 16(z + 1)$. This suggests the change of coordinates

$$\tilde{x} = x, \ \tilde{y} = y - 2, \ \tilde{z} = z + 1,$$

which yields an equation of the canonical type for the intersection parabola namely, $\tilde{x}^2 = 16\tilde{z}$. Now, it follows easily that the focus is

$F(\tilde{x} = 0, \tilde{y} = 0, \tilde{z} = 8)$ i.e., $F(x = 0, y = 2, z = 7)$.

3.2.19. Clearly, we must do exactly the same computations as we did for equations (3.2.3) and (3.2.4) but, with the replacement of y^2 by $y^2 + z^2$. Hence, the locus is an ellipsoid (two-sheeted hyperboloid) of revolution.

3.3.1. If the contact point of the required tangent is $P(x_0, y_0)$, the equation of the tangent put under the normal form is

$$\frac{16x_0x - 9y_0y - 144}{\sqrt{256x_0^2 + 81y_0^2}} = 0.$$

The distance from the origin to this line is $144/\sqrt{256x_0^2 + 81y_0^2}$. The distance from the focal point $F(5, 0)$ to the same line is $|80x_0 - 144|/\sqrt{256x_0^2 + 81y_0^2}$. These distances are equal iff $x_0 = 18/5$, $y_0 = \pm 4\sqrt{11}/5$, which defines P. The symmetric points of these P with respect to O also yield solutions of the problem.

3.3.2. Take the point $P((3/20)t^2, t)$ on the parabola. After polarization, we may write the equation of the tangent to the parabola at P as

$$(*) \qquad x = \frac{3t}{20}(2y - t).$$

Now, we will ask this line to cut the ellipse at a double point. If (*) is inserted in the equation, we get the intersection equation

$$y^2(4t^2 + 100) - 4t^3y + (t^4 - 2000) = 0.$$

This equation has equal solutions iff $t^4 - 80t^2 - 2000 = 0$, which has the real solutions $t = \pm 10$. For these values of, t (*) yields the required common tangents.

3.3.3. Since the conic, say Γ, passes through the origin, its equation is of the form

$$a_{11}x^2 + a_{22}y^2 + 2a_{12}xy + 2a_{10}x + 2a_{20}y = 0.$$

Γ passes through $(0, -1)$ iff $a_{22} - 2a_{20} = 0$, and then by a polarization, the tangent at $(0, -1)$ is

$$(a_{10} - a_{12})x + (a_{20} - a_{22})y - a_{20} = 0.$$

Since we want this line to be $x - y - 1 = 0$, we must have $a_{20} \neq 0$. Hence, after dividing the equation of Γ by a_{20}, we may assume that $a_{20} = 1$. Accordingly, $a_{22} = 2$, and the tangent is the desired line iff $a_{10} - a_{12} = 1$.

Now, if the obtained equalities are used, the point $(1, -2)$ belongs to Γ iff $a_{11} - 2a_{10} + 8 = 0$, and then the tangent at this point has the equation

$$(a_{11} - 4)x + a_{11}y + (a_{11} + 4) = 0.$$

This equation defines the line $4x + 3y + 2 = 0$ iff $a_{11} = -12$.

Thus, finally, the equation of Γ is

$$12x^2 - 2y^2 + 6xy + 4x - 2y = 0.$$

3.3.4. It is geometrically natural to look for a circumscribed square with vertices on the axes of the ellipse namely, $P(\lambda, 0)$, $Q(0, \lambda)$, $R(-\lambda, 0)$, $S(0, -\lambda)$. If we ask (PQ) $x + y - \lambda = 0$ to be tangent to the ellipse, we get $\lambda = 3$ (via the fact that the intersection equation has equal solutions). Hence, the sides of the required square are $x + y = \pm 3$, $x - y = \pm 3$.

As a matter of fact this is the only circumscribed square. Indeed, if $PQRS$ is a circumscribed square of an ellipse, the contact points M, N of PQ and RS with the ellipse are the endpoints of the diameter conjugated with the direction of PQ. Hence, they are symmetric with respect to the center O of the ellipse. The same holds for the contact points M', N' of QR, SP, and the direction QR. Accordingly, Q and S are symmetric with respect to O, and so are P and R. Therefore, the diagonals of the square pass through O. Then, using the result of Problem 3.3.21, QS must be the diametral line conjugated to the direction of MM', and PR is conjugated to MN'. But, from the definition of conjugation, it is clear that the directions of MM', $M'N'$ are conjugated too. Therefore the diagonals of the square are conjugated and orthogonal, and they must be along the symmetry axes of the ellipse.

3.3.5. If the hyperbola is $x^2/a^2 - y^2/b^2 - 1 = 0$, and $M(x_0, y_0)$ satisfies its equation, the tangent is

$$\frac{xx_0}{a^2} - \frac{yy_0}{b^2} - 1 = 0,$$

and it cuts the asymptotes $y = \pm(b/a)x$ at

$$x_{1,2} = \frac{a^2 b}{bx_0 \mp ay_0}, \quad y_{1,2} = \frac{\pm ab^2}{bx_0 \mp ay_0}.$$

(Here, the solution of index 1 is for the first sign and 2 for the second sign.) It follows easily that $x_1 + x_2 = 2x_0$, $y_1 + y_2 = 2y_0$.

3.3.6. Use a frame with origin at M_0. Then equation (3.3.6) of Γ becomes

$$(*) \qquad \bar{r} \cdot T\bar{r} + 2\bar{a} \cdot \bar{r} = 0.$$

Now, let $\bar{v}$, $\bar{w}$ be the unit vectors of $\vec{M_0A}$, $\vec{M_0B}$. The line M_0A has the equation $\bar{r} = \lambda\bar{v}$, and if this is inserted in (*) we get $\lambda(A)$, then

$$\bar{r}_A = \frac{2\bar{a} \cdot \bar{v}}{\bar{v} \cdot T\bar{v}} \bar{v}.$$

Similarly

$$\bar{r}_B = \frac{2\bar{a} \cdot \bar{w}}{\bar{w} \cdot T\bar{w}} \bar{w}.$$

The tangent of Γ at M_0 (as given by (3.3.25)) is $\bar{a} \cdot \bar{r} = 0$, therefore, the normal is $\bar{r} = t\bar{a}$. Then, the intersection point of the normal with the line AB is given by

$$(**) \qquad t\bar{a} = \alpha\bar{r}_A + \beta\bar{r}_B, \qquad \alpha + \beta = 1.$$

Inserting the known values of $\bar{r}_A$, $\bar{r}_B$ in (**), then scalarly multiplying (**) by $\bar{v}$ and $\bar{w}$, successively, while using $\bar{v}^2 = \bar{w}^2 = 1$, $\bar{v} \cdot \bar{w} = 0$, and $\alpha + \beta = 1$, we get $t(\bar{v} \cdot T\bar{v} + \bar{w} \cdot T\bar{w}) = 2$.

Now, if we assume that $\bar{v}(\cos\varphi, \sin\varphi)$, $\bar{w}(-\sin\varphi, \cos\varphi)$, and if we use the equivalent (3.3.1) of (*), we get

$$\bar{v} \cdot T\bar{v} + \bar{w} \cdot T\bar{w} = a_{11} + a_{22} = const.$$

Hence, $t = const.$, and the intersection point of AB with the normal is a fixed point, as required.

3.3.7. Since the two conics have the same focuses they have the same canonical frame, and we may write their equations as

$$\frac{x^2}{a^2} + \frac{y^2}{b^2} = 1, \quad \frac{x^2}{\alpha^2} - \frac{y^2}{\beta^2} = 1,$$

where $a^2 - b^2 = \alpha^2 + \beta^2$. The system of these two equations can be solved easily, and the intersection point in the first quadrant is

$$x = a\alpha\sqrt{\frac{b^2+\beta^2}{a^2\beta^2+b^2\alpha^2}}, \quad y = b\beta\sqrt{\frac{a^2-\alpha^2}{a^2\beta^2+b^2\alpha^2}}.$$

Now, we may use polarization to write down the equation of the tangents of the two conics at this intersection point, and it follows that the corresponding normals have the direction parameters $(\alpha/a, \beta/b)$, $(a/\alpha, -b/\beta)$, respectively. Hence, they are orthogonal.

3.3.8. If Γ is an ellipse or a hyperbola, we may write its equation as

$$\frac{x^2}{a^2} + \epsilon\frac{y^2}{b^2} - 1 = 0 \qquad (\epsilon = \pm 1).$$

The quadratic equations of the tangents from $M_0(x_0, y_0)$ to Γ is

$$\left(\frac{xx_0}{a^2} + \epsilon\frac{yy_0}{b^2} - 1\right)^2 - \left(\frac{x^2}{a^2} + \epsilon\frac{y^2}{b^2} - 1\right)\left(\frac{x_0^2}{a^2} + \epsilon\frac{y_0^2}{b^2} - 1\right) = 0,$$

which, after computations and since $\epsilon^2 = 1$, becomes

$$\epsilon b^2(x-x_0)^2 + a^2(y-y_0)^2 - (xy_0 - yx_0)^2 = 0.$$

By replacing $xy_0 - yx_0 = (x-x_0)y_0 - (y-y_0)x_0$, this equation becomes

$$(y-y_0)^2(a^2-x_0^2) + 2x_0y_0(y-y_0)(x-x_0) + (x-x_0)^2(\epsilon b^2 - y_0^2) = 0.$$

The slopes m_1, m_2 of the tangents are the solutions of the previous equation seen as a quadratic equation with respect to $(y-y_0)/(x-x_0)$. Hence, we want

$$m_1m_2 = \frac{\epsilon b^2 - y_0^2}{a^2 - x_0^2} = -1$$

i.e., the required locus of M_0 is the circle

$$x_0^2 + y_0^2 = a^2 + \epsilon b^2,$$

called the *Monge circle*.

If Γ is the parabola $y^2 - 2px = 0$, the quadratic equation of the tangents is

$$[yy_0 - p(x + x_0)]^2 - (y^2 - 2px)(y_0^2 - 2px_0) = 0,$$

and it is equivalent to

$$p(x - x_0)^2 - 2y_0(x - x_0)(y - y_0) + 2x_0(y - y_0)^2 = 0.$$

Hence, we must ask

$$m_1 m_2 = \frac{2x_0}{p} = -1,$$

and the locus is the directrix $x_0 = -(p/2)$.

3.3.9. Write the equation of Γ as $y^2-2px = 0$. Then $F(p/2, 0)$, (δ) $x = -p/2$. Take $M(x_0, y_0)$. The tangent of Γ at M is $yy_0 - px - px_0 = 0$, and $N(-p/2,\ p(2x_0 - p)/2y_0)$. Thus, the parallel by N to Ox is

$$(*) \qquad y = \frac{p}{2y_0}(2x_0 - p),$$

and we have to look for its intersection with

$$(MF) \quad \frac{x - p/2}{x_0 - p/2} = \frac{y}{y_0}.$$

For the intersection point, y is given by (*) and

$$x = \frac{p}{2} + \frac{p}{4y_0^2}(2x_0 - p)^2.$$

A comparison of this result and (*) shows that the required locus is

$$(**) \qquad x = \frac{p}{2} + \frac{y^2}{p}.$$

An easy change of coordinates shows that the locus is again a parabola.

Notice that we made no use of the condition $y_0^2 - 2px_0 = 0$. Hence, if we refer to the polar line of any point M instead of the tangent at M, the resulting point of the indicated construction also lies on the parabola.

3.3.10. Using the notation of (3.3.6) and (3.3.25), if the pencil of conics is

$$f_1(\bar{r}) + \lambda f_2(\bar{r}) = 0,$$

the polar lines of $M_0(\bar{r}_0)$ form the pencil of lines

$$f_1(\bar{r}; \bar{r}_0) + \lambda f_2(\bar{r}; \bar{r}_0) = 0.$$

3.3.11. If we write the equation of the conic under the general form (3.3.1), and ask this equation to be satisfied by $(0,0)$, $(1,0)$, $(0,1)$, we get $a_{00} = 0$, $2a_{10} = -a_{11}$, $2a_{20} = -a_{22}$. Then, for $x + y = 0$ to be a tangent of the conic we must also have $a_{11} = a_{22}$. Therefore, the family of conics is the sheaf

$$x^2 + y^2 - x - y + 2\lambda xy = 0.$$

These conics have the center at $x = y = 1/2(1 + \lambda)$. The characteristic equation is

$$\begin{vmatrix} 1-s & \lambda \\ \lambda & 1-s \end{vmatrix} = 0,$$

with solutions $s_1 = 1 - \lambda$, $s_2 = 1 + \lambda$, and the corresponding eigenvectors $(1, -1)$, $(1, 1)$.

Thus, the axes are

$$x + y = \frac{1}{1+\lambda}, \quad x - y = 0,$$

if $\lambda \neq -1$, and $x - y = 0$ only, if $\lambda = -1$.

3.3.12. With the general equation (3.3.1), and if $A(a, 0)$, we get $a_{00} = 0$, $2a_{10} = -a_{11}$, $a_{20} = 0$. Then, if m_1, m_2 are the slopes of the asymptotic directions, they are the solutions of

$$a_{11} + 2a_{12}\frac{y}{x} + a_{22}\left(\frac{y}{x}\right)^2 = 0,$$

and $m_1 m_2 = -1$ yields $a_{22} = -a_{11}$. Therefore, we remain with the family of conics

$$x^2 - y^2 + ax + 2\lambda xy = 0,$$

where $a = const.$, and λ is a parameter. The center is given by the equations

$$x + \lambda y - \frac{a}{2} = 0, \quad \lambda x - y = 0,$$

and the elimination of λ shows that the locus is the circle

$$x^2 + y^2 - \frac{ax}{2} = 0.$$

3.3.13. From the geometric definition of a conjugate diameter as a locus of midpoints it follows that the symmetry axes of the rectangle are conjugated diameters. Since they are also orthogonal, they coincide with the axes of the conic.

3.3.14. As a consequence of Proposition 3.3.15, the required equation is

$$\alpha(x - x_0)^2 + 2\beta(x - x_0)(y - y_0) + \gamma(y - y_0)^2 + \lambda = 0,$$

where α, β, γ, λ are arbitrary constants.

3.3.15. For $x^2/a^2 - y^2/b^2 = 1$ the polar line of $M(\alpha, \pm(b/a)\alpha)$ obtained by polarization has the slope $\pm b/a$.

3.3.16. The center of Γ is $C(-1/5, 3/5)$, and the translation $x = \tilde{x} - 1/5$, $y = \tilde{y} + 3/5$ puts the origin at C. The new equation of Γ will be

$$2\tilde{x}^2 - 12\tilde{x}\tilde{y} - 7\tilde{y}^2 + 1 = 0.$$

As in the usual case of a quadratic equation, the group of quadratic terms has a decomposition which transforms the equation of Γ into

$$2(\tilde{x} - \frac{6 + 5\sqrt{2}}{2}\tilde{y})(\tilde{x} - \frac{6 - 5\sqrt{2}}{2}\tilde{y}) + 1 = 0.$$

Now, if we go to the affine coordinates

$$x' = \tilde{x} - \frac{6 + 5\sqrt{2}}{2}\tilde{y}, \quad y' = \tilde{x} - \frac{6 - 5\sqrt{2}}{2}\tilde{y},$$

Γ becomes $x'y' = -1/2$, and we are done.

3.3.17. The required equation is

$$(ax + by + c)(a'x + b'y + c') = k = const.,$$

since by an obvious affine change of coordinates this reduces to $x'y' = k$.

3.3.18. If only affine notions come into the play, the simplest way to deal with the asymptotes is by using an affine frame with the axes along the asymptotes. Then $(\mathcal{H})$ $xy = k = const.$ Then, if $A(x_1, y_1)$, $C(x_2, y_2)$, where $x_1y_1 = k$, $x_2y_2 = k$, since the sides of the parallelogram are parallel to the axes, we must have $B(x_1, y_2)$, $D(x_2, y_1)$. Now, it follows easily that $O(0,0)$ satisfies the equation of BD.

3.3.19. The required point is the intersection of the given line with the conjugated diameter of the direction $(-3, 1)$ of that line. The equation of this diameter as given by (3.3.34) is $4x + 3y - 3 = 0$, and its intersection with $x + 3y - 12 = 0$ is $(-3, 5)$.

3.3.20. The center of the conic has the equations

$$2x - 3y - 6 = 0, \quad -3x + 9y + 7 = 0.$$

Hence, the center is $C(8/3, 4/9)$. If m is the slope of the required diameter, $(1, m)$ is a vector along it, and the conjugated direction has the direction parameters $(2 - 3m, -3 + 9m)$. The corresponding slope is $m' = (9m - 3)/(2 - 3m)$, and the angle condition is

$$\frac{m' - m}{1 + mm'} = \pm 1.$$

The two corresponding second order equations have the solutions $(5/3, 1/2)$ and $(1/4, -1/3)$, and we can write down the equations of the four lines which have these slopes and pass through C.

3.3.21. AB is the polar line of M_0, and we may write it under the form (3.3.25):

$$\bar{r} \cdot T\bar{r}_0 + \bar{a} \cdot (\bar{r} + \bar{r}_0) + \alpha = 0.$$

Then, if $\bar{m}$ is a vector along AB, we must have $\bar{m} \cdot (T\bar{r}_0 + \bar{a}) = 0$. On the other hand, if $\bar{\rho}_0$ is the radius vector of C, we have $T\bar{\rho}_0 + \bar{a} = \bar{0}$. It follows that

$$\bar{m} \cdot T(\vec{C_0M_0}) = \bar{m} \cdot (T\bar{r}_0 - T\bar{\rho}_0) = \bar{m} \cdot (T\bar{r}_0 + \bar{a}) = 0,$$

which means that C_0M is the conjugated diameter of AB, and it contains the midpoint of AB.

3.3.22. The equations of the center are

$$5x + 12y + 2 = 0, \quad 12x - 2y = 0,$$

with the solution $C(-2/77, -12/77)$.

The secular equation is

$$\begin{vmatrix} 5-s & 12 \\ 12 & -2-s \end{vmatrix} = s^2 - 3s - 154 = 0,$$

with the solutions $s_1 = 14$, $s_2 = -11$. Hence, the first axis has the direction parameters v_1, v_2 such that $-9v_1 + 12v_2 = 0$ i.e., its slope is $m_1 = v_2/v_1 = 3/4$. The axis of this slope through C has the equation $33x - 44y - 6 = 0$. Similarly, for s_2, we have $16v_1 + 12v_2 = 0$, $m_2 = v_2/v_1 = -4/3$, and the axis is $28x + 21y + 4 = 0$.

3.3.23. Check that the point is on the ellipsoid. By polarization we get the tangent plane $2x + 2y - 4z - 1 = 0$, and the normal is

$$\frac{x-2}{2} = \frac{y-1}{2} = \frac{z+1/2}{-4}.$$

3.3.24. The tangent plane at $M_0(x_0, y_0, z_0)$ is

$$\frac{xx_0}{21} + \frac{yy_0}{6} + \frac{zz_0}{4} = 1,$$

and it will be parallel to the given plane if $x_0/42 = y_0/12 = z_0/(-12)$. Therefore, $x_0 = -(7/2)z_0$, $y_0 = z_0$, and, since this point is on the given ellipsoid, we find $z_0 = \pm 1$, etc.

3.3.25. The problem is equivalent to that of finding the planes of the pencil through the given line, which have the pole on the quadric. The pencil of planes is

$$2x - 3z - 6 + \lambda(x + 2y) = 0,$$

and a polar plane is of the form

$$\frac{xx_0}{36} + \frac{yy_0}{9} - \frac{zz_0}{4} - 1 = 0.$$

The two planes coincide if the coefficients are proportional, which gives

$$x_0 = 6\lambda + 12,\ y_0 = 3\lambda,\ z_0 = 2.$$

Now, if we ask this point to be on the quadric, we get $\lambda = -1$. Hence, there exists only one plane as required, and it has the equation $x - 2y - 3z - 6 = 0$.

As a matter of fact, we were expecting two planes of this kind. The reason that we have only one is that the given line is tangent to the quadric, as it is easy to check.

3.3.26. We assume that the quadric has the equation (3.3.2). Since the quadric is cut by the plane $y = 0$ following the circle

$$(x - r)^2 + (y - r)^2 - r^2 = 0,\ \ y = 0,$$

our (3.3.2) must take the form

$$x^2 + a_{22}y^2 + z^2 + 2a_{12}xy + 2a_{23}yz - 2rx + 2a_{20}y - 2rz + r^2 = 0.$$

Since it is cut by $x = 0$ following a similar circle, we must have $a_{22} = 1$, $a_{23} = 0$, $a_{20} = -r$. Finally, the intersection with $z = 0$ is a parabola if it has no center, which happens iff $a_{12} = -1$. (For $a_{12} = 1$, we just get the line $x + y - r = 0$, twice.) Therefore, the required quadric is

$$x^2 + y^2 + z^2 - 2xy - 2rx - 2ry - 2rz + r^2 = 0.$$

3.3.27. If we choose the origin at the vertex of the cone, in view of Proposition 2.3.9 the cone has an equation of the form

$$(*)\qquad a_{11}x^2 + a_{22}y^2 + a_{33}z^2 + 2a_{12}xy + 2a_{13}xz + 2a_{23}yz = 0.$$

A generator is

$$x_0 = \alpha t,\ \ y_0 = \beta t,\ \ z_0 = \gamma t,$$

where α, β, γ are constants, and t is a parameter. The equation of the tangent plane given by polarization with (x_0, y_0, z_0) does not depend on t.

For the second part of the problem, let us use an affine frame with Ox along g_1, Oy along g_2, and Oz along the intersection of π_{g_1}, π_{g_2}. Then $a_{11} = a_{22} = 0$,

$$(\pi_{g_1})\ \ a_{12}y + a_{13}z = 0,\ \ (\pi_{g_2})\ \ a_{12}x + a_{23}z = 0,$$

and the planes intersect following Oz iff $a_{13} = a_{23} = 0$. Therefore, (*) reduces to

$$z^2 + \lambda xy = 0 \qquad (\lambda = 2a_{12}/a_{33}).$$

Accordingly, the diametral plane conjugated with $\bar{v}(0,0,1)$ is $z = 0$.

3.3.28. The last condition shows that the equation of the quadric must be of the form

$$x^2 - 4xy - 1 + a_{33}z^2 + 2a_{13}xz + 2a_{23}yz + 2a_{30}z = 0.$$

The equations of the center are

$$x - 2y + a_{13}z = 0, \; -2x + a_{23}z = 0, \; a_{13}x + a_{23}y + a_{33}z + a_{30} = 0,$$

and, since $C(0,0,1)$, we must have

$$a_{13} = 0, \; a_{23} = 0, \; a_{33} + a_{30} = 0.$$

After that, if we ask $M(2,0,-1)$ to be on the quadric, we get $a_{33} = -a_{30} = -1$, and the required equation is

$$x^2 - z^2 - 4xy + 2z - 1 = 0.$$

3.3.29. $4(x-1)^2 + 6(y+1)^2 + 4(z-3)^2 + 4(x-1)(z-3) = 0$.

3.3.30. By the usual procedure we find that the center of the quadric is $(1,1,1)$. Thus, the diameter through the origin is $x = y = z$ of direction $\bar{v}(1,1,1)$. The conjugated diametral plane is $\bar{r} \cdot T\bar{v} + \bar{a} \cdot \bar{v} = 0$ i.e., $x - 2y + 1 = 0$.

3.3.31. Using formulas (3.3.20), (3.3.21), we have the two families of rectilinear generators

$$\frac{x}{4} - \frac{y}{2} = t, \; \frac{x}{4} + \frac{y}{2} = \frac{z}{t},$$

$$\frac{x}{4} + \frac{y}{2} = t, \; \frac{x}{4} - \frac{y}{2} = \frac{z}{t}.$$

The direction of the first family is $(2,1,t)$, and it belongs to the given plane if it is orthogonal to $(3,2,-4)$. The result is $t = 2$, with one required generator

$$x - 2y = 8, \; x + 2y - 2z = 0.$$

Similarly, the direction of the lines of the second family is $(2,-1,t)$, and we get $t = 1$, with the generator

$$x + 2y = 4, \quad x - 2y - 4z = 0.$$

3.3.32. With the usual method we find that the quadric has the center $C(1,-1,1)$. The secular equation

$$\begin{vmatrix} 1-s & -3 & -1 \\ -3 & 1-s & 1 \\ -1 & 1 & 5-s \end{vmatrix} = -(s^3 - 7s^2 + 36) = 0$$

has the solutions $s_1 = -2$, $s_2 = 3$, $s_3 = 6$. The corresponding basic eigenvectors are $(1,1,0)$, $(1,-1,1)$, $(-1,1,2)$. Hence, the symmetry axes are

$$\frac{x-1}{1} = \frac{y+1}{1} = \frac{z-1}{0}; \quad \frac{x-1}{1} = \frac{y+1}{-1} = \frac{z-1}{1};$$

$$\frac{x-1}{-1} = \frac{y+1}{1} = \frac{z-1}{2}.$$

3.3.33. The center of this quadric is $C(1,1,-1)$. The solution of its secular equation are $s_1 = s_2 = -3$, $s_3 = 6$. For s_3 the basic eigenvector is $(2,1,-2)$, and the plane through C perpendicular to this eigenvector i.e.,

$$(*) \qquad 2(x-1) + (y-1) - 2(z+1) = 0$$

is a symmetry plane.

For $s_1 = s_2 = -3$ the characteristic system reduces to the single equation

$$2v_1 + v_2 - 2v_3 = 0,$$

and we get infinitely many symmetry planes of equations

$$a(x-1) + 2(b-a)(y-1) + b(z+1) = 0 \quad (a, b \in \mathbf{R}).$$

These planes are orthogonal to (*).

Hence, the intersection conic of the quadric with planes which are parallel to (*) have infinitely many symmetry axes, and they are circles with the center on the symmetry axis of s_3. This shows that we have a quadric of revolution.

3.3.34. We use again the general equation (3.3.2). The fact that the quadric contains any point of Oz yields $a_{33} = a_{30} = a_{00} = 0$. Then, the tangent plane at O is $x = 0$ iff $a_{20} = 0$, and Ox, Oy are asymptotic directions iff $a_{11} = a_{22} = 0$. If we polarize the remaining equation at $(1,1,1)$, and ask the obtained plane to have coefficients proportional to those of the given plane, we get

$$a_{11} = a_{13},\ a_{23} = -3a_{13},\ a_{10} = -4a_{13}.$$

Hence, the required quadric has the equation

$$xy + xz - 3yz - 4x = 0.$$

3.3.35. Use an affine frame with origin at C, xOy in π, and Oz the conjugated diameter. Then, the general equation (3.3.2) becomes

$$a_{11}x^2 + a_{22}y^2 + a_{33}z^2 + 2a_{12}xy + 2a_{13}xz + 2a_{23}yz + a_{00} = 0,$$

and, since the conjugated plane of $(0,0,1)$ is $z = 0$, we also have $a_{13} = a_{23} = 0$. Now, the intersection γ of Γ with $z = k = const.$ has the center $C(c_1, c_2, k)$ defined by the equations

$$a_{11}c_1 + a_{12}c_2 = 0,\ a_{21}c_1 + a_{22}c_2 = 0$$

i.e., $c_1 = c_2 = 0$ (because $\delta(\Gamma) \neq 0$ implies $\delta(\gamma) \neq 0$). Hence, the centers of γ belong to Oz.

3.3.36. If Γ would be a conic, and we look at tangent lines instead of planes, the same result as in the problem is obvious since the contact points may be seen as midpoints of the chord determined by the tangent line. If Γ is a quadric, we use the previous result for all the intersection conics of Γ by planes through the line which joins the contact points or through the given diameter, for the second assertion. (Another possibility is to use the result of Problem 3.3.35.)

3.3.37. First, we look for the planes which cut the paraboloid by circles. It is clear that a plane which is parallel to Oz cannot do so. Hence, we may put the equation of the plane in the form

$$z = \alpha x + \beta y + \gamma.$$

The normal of this plane has the direction $(\alpha, \beta, -1)$, and it is easy to find an orthonormal basis of the plane, say

$$\bar{u}(-\beta/\theta, \alpha/\theta, 0),\ \bar{v}(\frac{\alpha}{\theta\sqrt{1+\theta^2}}, \frac{\beta}{\theta\sqrt{1+\theta^2}}, \frac{\theta}{\sqrt{1+\theta^2}}),$$

where $\theta^2 = \alpha^2 + \beta^2$. If we use this basis, and the point $(0, 0, \gamma)$ of the plane, and write the parametric equations of our plane as

$$x = -\lambda\frac{\beta}{\theta} + \mu\frac{\alpha}{\theta\sqrt{1+\theta^2}},$$

$$y = \lambda\frac{\alpha}{\theta} + \mu\frac{\beta}{\theta\sqrt{1+\theta^2}},$$

$$z = \gamma + \mu\frac{\theta}{\sqrt{1+\theta^2}},$$

λ, μ are internal orthogonal coordinates of the plane.

If the above mentioned parametric expressions are inserted in the equation of the paraboloid, and if it is required that the result does not contain $\lambda\mu$-term, we get either $\alpha = 0$ or $\beta = 0$. Furthermore, if we ask that the coefficients of λ^2 and μ^2 of the intersection equation be equal, we get a real solution only for $\alpha = 0$, and the solution is $\beta = \pm 1$. ($\beta = 0$ leads to imaginary values of α.)

Accordingly, there are two pencils of parallel planes which cut the paraboloid by circles, and they have the equations

$$z = \pm y + \gamma \qquad (\gamma = const.).$$

Finally, we have to find those planes of these pencils for which the pole belongs to the paraboloid. (See the solution of Problem 3.3.25, for instance.) This happens iff $\gamma = 1/2$. Therefore, the required planes are $z = \pm y + 1/2$, and their contact points (called *cyclic points* of the quadric) are $(0, \pm 1, 1/2)$.

3.4.1. Use the three methods which were used for Example 3.4.3. The result is that 1) is an ellipse, 2) a hyperbola and 3) a parabola.

3.4.2. Again, work similarly as in Example 3.4.3. The results are: 1) one-sheeted hyperboloid; 2) two-sheeted hyperboloid; 3) ellipsoid; 4) hyperbolic paraboloid; 5) elliptic cylinder.

3.4.3. The quadrics have $\delta = -\Delta = \alpha^2 - 2$. For $\alpha = \pm\sqrt{2}$, the Gauss method shows that we have a hyperbolic cylinder. In all cases, the cone of asymptotic directions at the origin is cut by $z = 0$ following the real lines $z = 0$, $x \pm y = 0$. Hence, if we also look at the sign of Δ, we see that the quadric is a one-sheeted hyperboloid for $-\sqrt{2} < \alpha < \sqrt{2}$, and a two-sheeted hyperboloid if either $\alpha < -\sqrt{2}$ or $\alpha > \sqrt{2}$.

3.4.4. We have $\delta = -\alpha^2$, $\Delta = 5\alpha^2 - 36$. If $\alpha = \pm 6/\sqrt{5}$, the quadric is degenerated with a unique symmetry center, and the Gauss method shows that it is a hyperbolic cone. For $\alpha = 0$, the quadric is an elliptic paraboloid. The cone of asymptotic directions at the origin is cut by $z = 0$ following a real line: $z = 0$, $y = 0$. Accordingly, for either $\alpha < -6/\sqrt{5}$ or $\alpha > 6/\sqrt{5}$ we have a one-sheeted hyperboloid, and for $-6/\sqrt{5} < \alpha < 6/\sqrt{5}$, $\alpha \neq 0$, we have a two-sheeted hyperboloid.

3.4.5. We have $\delta = \alpha - 4$, $\Delta = -8(\alpha - 4) - (\beta + 6)^2$. If $\Delta \neq 0$, the conic is an ellipse for $\alpha > 4$, a parabola for $\alpha = 4$, and a hyperbola for $\alpha < 4$. If $\Delta = 0$ but $\beta \neq -6$, $\alpha - 4 = \delta < 0$, and the conic is a pair of real, crossing lines. If $\Delta = 0$, $\beta = -6$, then $\alpha = 4$, and we have two real, parallel lines.

3.4.6. i) We have $\delta = 1 - \alpha^2 - \beta^2$, $\Delta = (\alpha + 1)(2\alpha^2 + 2\beta^2 - 1)$. Therefore, if $P(\alpha, \beta)$ is outside the circle of center $O(0,0)$ and radius 1, but not on the line $\alpha = -1$, the conic is a hyperbola. If P is on this circle, but not on the line $\alpha = -1$, the conic is a parabola. If P is inside this circle, but not on the circle of center O and radius $1/\sqrt{2}$, the conic is an ellipse. If P is on the line $\alpha = -1$, the conic is degenerated. Namely, if $\beta > 1$ we have two real, crossing lines, if $\beta < 1$ we have two imaginary crossing lines, and if $\beta = 1$ two coincident lines. If P is on the circle $\alpha^2 + \beta^2 = 1/\sqrt{2}$, we have two imaginary lines.

ii) $\delta = \alpha\beta$, $\Delta = \beta[\alpha(\beta - \alpha) - 1]$. Therefore, in the second and third quadrants, but not on the axes of the (α, β)-plane, the conic is a hyperbola. In the first and the third quadrant, not on the axes or on the hyperbola $\alpha(\beta - \alpha) = 1$, the conic is an ellipse. On $\alpha = 0$, $\beta \neq 0$, we have a parabola. On $\beta = 0$ and on the hyperbola $\alpha(\beta - \alpha) = 1$, the conic is a pair of real, crossing, straight lines.

4.1.1.

$$(\varphi^{-1}\circ\psi^{-1})\circ(\psi\circ\varphi)=\varphi^{-1}\circ(\psi^{-1}\circ\psi)\circ\varphi$$
$$=\varphi^{-1}\circ id_{\mathcal{F}}\circ\varphi=\varphi^{-1}\circ\varphi=id_{\mathcal{F}}.$$

4.1.2. No, since, for instance, $\varphi:\mathcal{F}\longrightarrow\mathcal{F}'$ and $\psi:\mathcal{F}_1\longrightarrow\mathcal{F}_1'$ cannot be composed if $\mathcal{F}'$ is not included in $\mathcal{F}_1$.

4.2.1. Let us consider the translations of vectors $\bar{a}$, $\bar{b}$:

$$(\tau_1)\ x_i'=x_i+a_i,\quad (\tau_2)\ x_i'=x_i+b_i.$$

Then $\tau_2\circ\tau_1$ is

$$x_i'=(x_i+a_i)+b_i=x_i+(a_i+b_i),$$

and this is the translation of vector $\bar{a}+\bar{b}$. Then, say τ_1^{-1} is $x_i=x_i'-a_i$, which is the translation of vector $-\bar{a}$.

4.2.2. Take O as origin, and the homotheties

$$(h_1)\ x_i'=k_1x_i,\quad (h_2)\ x_i'=k_2x_i.$$

Then $h_2\circ h_1$ is $x_i'=k_2(k_1x_i)=(k_2k_1)x_i$, and h_1^{-1} is $x_i=k_1^{-1}x_i'$.

4.2.3. If the two trapeziums are $ABCD$, $A'B'C'D'$, with the bases AB, CD; $A'B'$, $C'D'$, respectively, the points $M=AD\cap BC$, $M'=A'D'\cap B'C'$ correspond each to the other by the same affine transformation. Since the transformation preserves simple ratios we must have $AB/CD=A'B'/C'D'$. Conversely, if the last condition holds, the affine transformation which sends (C,D,M) to (C',D',M') also sends A to A' and B to B'.

4.2.4. These affine transformations must also keep fixed the pair of asymptotes. Hence, we either have

$$x'=kx,\quad y'=\frac{1}{k}y$$

or

$$x'=ky,\quad y'=\frac{1}{k}x.$$

4.2.5. i) Since all $AA' \| Oy$, the equations of the transformation must have the form

$$x' = x \quad y' = \alpha x + \beta y + \gamma.$$

But, we also want $y = 0$ to imply $y' = 0$ (then $(x, 0)$ are fixed points). Hence, $\alpha = 0$, $\gamma = 0$, and the answer is $x' = x$, $y' = \beta y$.

ii) The fixed points of the transformation are given by

$$x = 4x + y - 5, \quad y = 6x + 3y - 10.$$

These two conditions reduce to the single independent condition $3x + y - 5 = 0$. Hence, there exists a line of fixed points. On the other hand, a line which joins a point with its image has the slope $(y'-y)/(x'-x) = 2 = const.$ Hence, the transformation is an affine perspective.

4.2.6. As told in the hint given with the problem, if C is a circle, and φ is a transformation, $\varphi(C)$ is an ellipse, in view of the affine classification theorem of conics. The axes of this ellipse may be seen as a pair of conjugated diameters. Since this latter notion has an affine character, the axes are the image by φ of a pair of conjugated, hence orthogonal diameters of C.

4.2.7. We may use the method of the previous Problem 4.2.6, for instance. The circle

$$(x' - 5)^2 + (y' + 7)^2 = 1$$

is the image by φ of the ellipse

$$13x^2 - 24xy + 45y^2 = 1.$$

The latter has the secular equation $s^2 - 58s + 441 = 0$, with the solutions $s_1 = 49$, $s_2 = 9$, and with the principal vectors $\bar{v}_1(-1, 3)$, $\bar{v}_2(3, 1)$. Clearly, these are also principal directions of the given affine transformation. (Check with the equations of the latter, if you still need.)

4.2.8. If we make an affine transformation which sends the ellipse to a circle (which is always possible, obviously), the circumscribed parallelogram transforms into a square circumscribed to the circle. The result holds for the diagonals of this square, because they are orthogonal. Since the result has an affine character, it also holds for the original parallelogram. And, if the latter is a rhombus, its diagonals are conjugated and orthogonal, which means that they are situated on the symmetry axes of Γ.

4.2.9. If the transformation is (4.2.2) i.e.,

$$x'_j = \sum_{i=1}^{3} b_{ij}x_j + \beta_j,$$

it transforms the vectors by the rule

$$v'_j = \sum_{i=1}^{3} b_{ij}v_j.$$

The orthogonality condition $\sum_{j=1}^{3} v_j v'_j = 0$ holds iff $b_{ij} + b_{ji} = 0$. (In other words, the matrix $B = (b_{ij})$ of the transformation is skew symmetric.)

4.3.1. Make $x' = x$, $y' = y$ in the equations of the motion. The obtained system of equations has a unique solution if $\varphi \neq 0$, and no solutions if $\varphi = 0$ (in which case we have a translation). For $\varphi \neq 0$, using some trigonometry computations, one gets the fixed point

$$x = \frac{1}{2}(x_0 - y_0 \cot\frac{\varphi}{2}), \quad y = \frac{1}{2}(x_0 \cot\frac{\varphi}{2} + y_0).$$

4.3.2. If there exists a motion which sends A to B, B to C and C to A, we must have $AB = BC = CA$, and the triangle must be equilateral. This fact can be easily checked for the given points A, B, C. Then, it is clear that the required rotation exists, and it is the rotation of angle $\pi/3$ around the center of the circumscribed circle of the triangle. The equations of this rotation are like those of Problem 4.3.1 above, where $\varphi = \pi/3$ (since φ of those equations is the angle between a vector and its image i.e., just the rotation angle). Therefore,

$$x' = -\frac{1}{2}x + \frac{\sqrt{3}}{2}y + x_0, \quad y' = \frac{\sqrt{3}}{2}x - \frac{1}{2}y + y_0.$$

Finally, the condition that $(0,0) \longmapsto (b,0)$ yields $x_0 = b$, $y_0 = 0$.

4.3.3. Choose the axes so that $A(0,0)$, $B(a,0)$, $C(a,a)$, $D(0,a)$, and write down the equations of the rotation:

$$x' = x\cos\varphi - y\sin\varphi, \quad y' = x\sin\varphi + y\cos\varphi.$$

Then: $B'(a\cos\varphi, a\sin\varphi)$, $D'(-a\sin\varphi, a\cos\varphi)$, $C'(a(\cos\varphi - \sin\varphi), a(\cos\varphi + \sin\varphi))$. Furthermore:

$$(BB')\ \frac{x-a}{\cos\varphi - 1} = \frac{y}{\sin\varphi}, \quad (DD')\ \frac{x}{-\sin\varphi} = \frac{y-a}{\cos\varphi - 1}.$$

It is easy to find the locus of the intersection point of these two lines. The first equation is equivalent to

$$\frac{x-a}{y} = -\tan\frac{\varphi}{2},$$

and the second to

$$\frac{y-a}{x} = \tan\frac{\varphi}{2}.$$

Hence, the locus is the circle

$$x^2 + y^2 - ax - ay = 0.$$

The same equations yield the values of the coordinates of $BB' \cap DD'$:

$$x = a\cos^2\frac{\varphi}{2}\left(\tan\frac{\varphi}{2} - 1\right), \quad y = a\cos^2\frac{\varphi}{2}\left(\tan\frac{\varphi}{2} + 1\right),$$

and it remains to check that these coordinates satisfy the equation of CC' namely,

$$(CC')\ \frac{x-a}{\cos\varphi - \sin\varphi - 1} = \frac{y-a}{\cos\varphi + \sin\varphi - 1},$$

or, equivalently:

$$\frac{x-a}{\cos\frac{\varphi}{2} + \sin\frac{\varphi}{2}} + \frac{y-a}{\cos\frac{\varphi}{2} - \sin\frac{\varphi}{2}} = 0.$$

This follows by trigonometric calculations.

4.3.4. The induced linear transformation of vectors is

$$v_1' = v_1 \cos\theta + v_2 \sin\theta, \quad v_2' = v_1 \sin\theta - v_2 \cos\theta,$$

and the determinant of its coefficients is -1; hence, the transformation is an indirect isometry.

The axis of the required symmetry is a fixed line, necessarily. Hence, it has the direction of a fixed vector. (Remember that the length of the vectors is preserved.) This vector is a solution of

$$v_1(1 - \cos\theta) - v_2 \sin\theta = 0.$$

(The second condition is proportional to this one.) If we also ask ${v_1}^2 + {v_2}^2 = 1$, we get $\bar{v}(\cos\frac{\theta}{2}, \sin\frac{\theta}{2})$.

A line d of direction $\bar{v}$ through a point $P(a, b)$ has parametric equations

$$x = a + t\cos\frac{\theta}{2}, \quad y = b + t\sin\frac{\theta}{2},$$

and the given transformation sends it to the line d' of equations

$$x = [a\cos\theta + b\sin\theta + x_0] + t\cos\frac{\theta}{2},$$

$$y = [a\sin\theta - b\cos\theta + y_0] + t\sin\frac{\theta}{2}.$$

And, $d' = d$ if we choose (a, b) such that

$$a\cos\theta + b\sin\theta + x_0 = a + \tau\cos\frac{\theta}{2},$$

$$a\sin\theta - b\cos\theta + y_0 = b + \tau\sin\frac{\theta}{2},$$

or, equivalently

$$a(\cos\theta - 1) + b\sin\theta = \tau\cos\frac{\theta}{2} - x_0,$$

$$a\sin\theta - b(\cos\theta + 1) = \tau\sin\frac{\theta}{2} - y_0.$$

Here, the coefficients of a, b are proportional hence, solutions exist if this proportionality extends to the right hand sides of the equation i.e.,

$$\frac{\tau\cos\frac{\theta}{2} - x_0}{\tau\sin\frac{\theta}{2} - y_0} = \frac{\cos\theta - 1}{\sin\theta} = -\frac{\sin\frac{\theta}{2}}{\cos\frac{\theta}{2}}.$$

This yields

$$\tau = x_0\cos\frac{\theta}{2} + y_0\sin\frac{\theta}{2},$$

and then

$$a(\cos\theta - 1) + b\sin\theta = \sin\frac{\theta}{2}(-x_0\sin\frac{\theta}{2} + y_0\cos\frac{\theta}{2}).$$

Thus, if we assume $\theta \neq 0$ (if $\theta = 0$ the transformation is the symmetry with respect to the axis $y = y_0/2$ followed by the translation of vector $(x_0, 0)$), the line of direction $\bar{v}$ and through $P(a = 0, b = (1/2)(y_0 - x_0\tan(\theta/2)))$ is the axis of the required symmetry. And, $\vec{PQ}$ where Q is the image of P is the required translation. Namely,

$$\vec{PQ}(x_0\cos^2\frac{\theta}{2} + y_0\sin\frac{\theta}{2}\cos\frac{\theta}{2}, x_0\sin\frac{\theta}{2}\cos\frac{\theta}{2} + y_0\sin^2\frac{\theta}{2}),$$

and the symmetry axis is

$$(x - \frac{x_0}{2})\sin\frac{\theta}{2} - (y - \frac{y_0}{2})\cos\frac{\theta}{2} = 0.$$

4.3.5. Take the origin O at $d_1 \cap d_2$, and the x-axis along d_1. Assume that d_2 is the axis which makes an oriented angle α with d_1. From formula (4.3.8), we know that the corresponding polar coordinates change as follows

$$(\rho, \theta) \xrightarrow{\sigma_1} (\rho, -\theta) \xrightarrow{\sigma_2} (\rho, \theta + 2\alpha),$$
$$(\rho, \theta) \xrightarrow{\sigma_2} (\rho, -\theta + 2\alpha) \xrightarrow{\sigma_1} (\rho, \theta - 2\alpha).$$

$\sigma_2 \circ \sigma_1 = \sigma_1 \circ \sigma_2$ if the two final angles differ by a full rotation i.e., $4\alpha = 2\pi$. That is $d_1 \perp d_2$.

4.3.6. The center of the cube must be a fixed point of any symmetry of the cube. If we denote by $2a$ the length of the edge, and choose the coordinates as indicated, it follows easily that the required symmetry planes are $x = a$, $y = a$, $z = a$, $x = y$, $y = z$, $z = x$. By formula (4.3.6), the symmetry with respect to the plane $z = a$ is $x' = x$, $y' = y$, $z' = -z + 2a$, the symmetry with respect to $x = a$ is $x' = -x + 2a$, $y' = y$, $z' = z$, and the symmetry with respect to $y = a$ is $x' = x$, $y' = -y + 2a$, $z' = z$. The plane $x = y$ passes through Oz, and through the bisector of the first quadrant of the xy-plane. Hence, the equations of the corresponding symmetry are $x' = y$, $y' = x$, $z' = z$, etc.

4.3.7. The composition of the three indicated rotations acts on the basis $\bar\imath, \bar\jmath, \bar k$ as follows

$$(\bar\imath, \bar\jmath, \bar k) \longmapsto (\bar\xi, \bar\eta, \bar k) \longmapsto (\bar\xi, \bar\eta' := \bar k' \times \bar\xi, \bar k') \longmapsto (\bar\imath', \bar\jmath' = \bar k' \times \bar\imath', \bar k'),$$

and this justifies the first assertion.

Analytically, the indicated rotations are, respectively, given by

$$\bar\xi = \bar\imath \cos\varphi + \bar\jmath \sin\varphi, \quad \bar\eta = -\bar\imath \sin\varphi + \bar\jmath \cos\varphi, \quad \bar k = \bar k,$$

$$\bar\xi = \bar\xi, \quad \bar\eta' = \bar\eta \cos\theta + \bar k \sin\theta, \quad \bar k' = -\bar\eta \sin\theta + \bar k \cos\theta,$$

$$\bar\imath' = \bar\xi \cos\psi + \bar\eta' \sin\psi, \quad \bar\jmath' = -\bar\xi \sin\psi + \bar\eta' \cos\psi, \quad \bar k' = \bar k'.$$

By eliminating $\bar\xi, \bar\eta, \bar\eta'$ from these equations, we get

$$\bar\imath' = (\cos\varphi\cos\psi - \sin\varphi\sin\psi\cos\theta)\bar\imath + (\sin\varphi\cos\psi + \cos\varphi\sin\psi\cos\theta)\bar\jmath + \sin\psi\sin\theta\bar k,$$

$$\bar\jmath' = -(\cos\varphi\sin\psi - \sin\varphi\cos\psi\cos\theta)\bar\imath - (\sin\varphi\sin\psi - \cos\varphi\cos\psi\cos\theta)\bar\jmath + \cos\psi\sin\theta\bar k,$$

$$\bar k' = \sin\varphi\sin\theta\bar\imath - \cos\varphi\sin\theta\bar\jmath + \cos\theta\bar k.$$

The obtained formulas define the linear transformation of vectors induced by the rotation. Hence, the equations of the rotation with respect to the first frame are

$$x' = (\cos\varphi\cos\psi - \sin\varphi\sin\psi\cos\theta)x - (\cos\varphi\sin\psi - \sin\varphi\cos\psi\cos\theta)y + (\sin\varphi\sin\theta)z,$$

$$y' = (\sin\varphi\cos\psi + \cos\varphi\sin\psi\cos\theta)x - (\sin\varphi\sin\psi - \cos\varphi\cos\psi\cos\theta)y - (\cos\varphi\sin\theta)z,$$

$$z' = (\sin\psi \sin\theta)x + (\cos\psi \sin\theta)y + (\cos\theta)z.$$

4.3.8. A computation shows that the induced vector transformation preserves the length of a vector, and it has determinant +1. Hence, we have a direct isometry with the origin as a fixed point. To find its rotation axis we proceed as in the proof of Proposition 4.3.5. We get that the direction of the axis is $(7, 3, 1)$.

5.1.1. We prove the result by using the basic incidence properties. The second assertion of 5 and 2, Proposition 5.1.3, show that there are well-defined planes α, β which contain (A, a) and (A, b), respectively, and $\alpha \neq \beta$ since a, b are skew lines. Then, 4 and 1, Proposition 5.1.3, yield an intersection line $d = \alpha \cap \beta$, and 6, Proposition 5.1.3, shows that d intersects both a and b.

5.1.2. Denote $AA' \cap BB' \cap CC' = O$, $A'A'' \cap B'B'' \cap C'C'' = O'$, $AA'' \cap BB'' \cap CC'' = O''$, $P = AB \cap A'B'$, $P' = A'B' \cap A''B''$, $P'' = AB \cap A''B''$, $Q = BC \cap B'C'$, $Q' = B'C' \cap B''C''$, $Q'' = BC \cap B''C''$, $R = AC \cap A'C'$, $R' = A'C' \cap A''C''$, $R'' = AC \cap A''C''$, $d =$ the common axis of perspective. Then, we must have $P = d \cap A'B' = P' = d \cap A''B'' = P''$, and $Q = Q' = Q''$, $R = R' = R''$, similarly. Hence, $\triangle AA'A''$, $\triangle BB'B''$ are in a perspective with center P, and with the axis containing O, O', O''.

The notion of perspective triangles is self-dual by plane duality. Hence, the plane dual result is: if three triangles of a plane are pairwise perspective, with the same center, the three perspective axes have a common point.

5.2.1. i) The affine, nonhomogeneous coordinates of the vertices were computed in Problem 1.3.3, and they were $A_1(0,0)$, $A_2(1,0)$, $A_3(0,1)$, $A_4(-2,2)$, $A_5(-3,2)$, $A_6(-2,1)$. Hence, the homogeneous coordinates are $A_1[0:0:1]$, $A_2[1:0:1]$, $A_3[0:1:1]$, $A_4[-2:2:1]$, $A_5[-3:2:1]$, $A_6[-2:1:1]$. The improper point of $A_1A_2 \| A_5A_4$ has the coordinates define by $\bar{e}_1$ i.e., $[1:0:0]$, the improper point of $A_2A_3 \| A_6A_5$ is defined by $\bar{e}_2 - \bar{e}_1$ i.e., $[-1:1:0]$, and the improper point of $A_3A_4 \| A_1A_6$ is defined by $-2\bar{e}_1 + \bar{e}_2$ i.e., $[-2:1:0]$.

ii) We write the nonhomogeneous equations of the sides, then use (5.2.1). The results are (A_1A_2) $x_2 = 0$; (A_1A_3) $x_1 + x_2 - x_0 = 0$; (A_3A_4) $x_1 + 2x_2 - 2x_0 = 0$; (A_4A_5) $x_2 - 2x_0 = 0$; (A_5A_6) $x_1 + x_2 + x_0 =$

0; (A_1A_6) $x_1 + 2x_2 = 0$.

5.2.2. This is the same cube as the one of Problem 1.3.4, with $a = 1$. Remember that the coordinates of its vertices A and A', on Ox, are $(\pm 1/\sqrt{2}, 0, 0)$, B and B', on Oy, are $(0, \pm 1/\sqrt{2}, 0)$, $\tilde{A}$ and $\tilde{A}'$, over A, A', are $(\pm 1/\sqrt{2}, 0, 1)$, and $\tilde{B}$ and $\tilde{B}'$, over B, B', are $(0, \pm 1/\sqrt{2}, 1)$. The required improper points are the unmarked directions of the vectors $\vec{AB}[-1:1:0:0]$, $\vec{AB'}[1:1:0:0]$, $\vec{A\tilde{A}}[0:0:1:0]$. These points are not collinear since the three vectors do not belong to the same plane.

5.2.3. Since xOy has the homogeneous equation $x_3 = 0$, the required projective coordinates are defined by

$$\rho y_1 = \alpha_{11}x_1 + \alpha_{12}x_2 + \alpha_{13}x_3 + \alpha_{10}x_0,$$
$$\rho y_2 = \alpha_{21}x_1 + \alpha_{22}x_2 + \alpha_{23}x_3 + \alpha_{20}x_0,$$
$$\rho y_3 = \alpha_{31}x_1 + \alpha_{32}x_2 + \alpha_{33}x_3 + \alpha_{30}x_0,$$
$$\rho y_0 = x_3,$$

where the equations are invertible. The latter condition amounts to

$$\begin{vmatrix} \alpha_{11} & \alpha_{12} & \alpha_{10} \\ \alpha_{21} & \alpha_{22} & \alpha_{20} \\ \alpha_{31} & \alpha_{32} & \alpha_{30} \end{vmatrix} \neq 0.$$

5.3.1. The parametric equations of the given line are

$$x = 2 + t, \quad y = 3 + 2t, \quad z = 5 + 6t,$$

where t is an affine coordinate on the line. Cutting with the coordinate planes gives $t_A = -5/6$, $t_B = -2$, $t_C = -3/2$, and, of course, $t_P = 0$. Accordingly,

$$(A, B; C, P) = \frac{t_C - t_A}{t_C - t_B} : \frac{t_P - t_A}{t_P - t_B} = \frac{16}{5}.$$

5.3.2. i) Let A_1, A_2, A_3, A_4 be intersection points of h_1, h_2, h_3, h_4, respectively, with another line d of the plane. Then the sinus theorem of trigonometry yields

$$\frac{A_1A_3}{\sin \widehat{h_1h_3}} = \frac{OA_1}{\sin \widehat{OA_3A_1}}, \quad \frac{A_2A_3}{\sin \widehat{h_2h_3}} = \frac{OA_2}{\sin \widehat{OA_3A_2}},$$

$$\frac{A_1A_4}{\sin\widehat{h_1h_4}} = \frac{OA_1}{\sin\widehat{OA_4A_1}}, \quad \frac{A_2A_4}{\sin\widehat{h_2h_4}} = \frac{OA_2}{\sin\widehat{OA_4A_2}}.$$

And, $(d_1, d_2; d_3, d_4) = (A_1, A_2; A_3, A_4)$ yields the result.

ii) The arc which corresponds to the side of the pentagon is of 72°, and the result of i) yields

$$(A_1A_2, A_1A_3; A_1A_4, A_1A_5) = \frac{\sin 72^\circ}{\sin 36^\circ} : \frac{\sin 108^\circ}{\sin 72^\circ}.$$

5.3.3. We cut the lines AB, AC, AD, AE with the improper line of the plane, and get $P_\infty[1 : -1 : 0]$, $Q_\infty[1 : 0 : 0]$, $R_\infty[1/2 : 1 : 0]$, $S_\infty[1/4 : 1 : 0]$, respectively. (Of course, the first two coordinates are those of $\vec{AB}$, $\vec{AC}$, $\vec{AD}$, $\vec{AE}$.) Then, the required cross ratio is equal to that of the corresponding points on the projective line $x_0 = 0$ i.e.,

$$(P_\infty, Q_\infty; R_\infty, S_\infty) = \frac{\begin{vmatrix} 1 & -1 \\ 1/2 & 1 \end{vmatrix}}{\begin{vmatrix} 1 & 0 \\ 1/2 & 1 \end{vmatrix}} : \frac{\begin{vmatrix} 1 & -1 \\ 1/4 & 1 \end{vmatrix}}{\begin{vmatrix} 1 & 0 \\ 1/4 & 1 \end{vmatrix}} = \frac{6}{5}.$$

5.3.4. Choose the origin at M. Then, $A(-a)$, $B(a)$, $C(c)$, $D(d)$, and $(A, B; C, D) = -1$ is equivalent to $a^2 = cd$. But, $a^2 = MA^2$, and $MN^2 - NC^2 = cd$. Hence, we have the required result.

5.3.5. Denote $E = d \cap d'$; we may assume that none of the given points is E since the result is trivial otherwise. Assume that $PQ \cap B'C = R_1$. Denote $G = AB' \cap A'C$, $H = AC' \cap A'C$. By consecutive projections from Q on AB', from A' on AB, and from C' on CB' we get

$$(C, H; B', R_1) = (G, A; B', P) = (C, A; E, B) = (C, H; B', R).$$

This shows that we must have $R_1 = R$. We advise the reader to give also a computational proof using affine, homogeneous coordinates with axes d, d'.

5.3.6. One has

$$(M, N; P, Q) = (d_1, d_2; OP, OQ) = -1,$$

and, since the line OP is a fixed line of the pencil defined by d_1, d_2, OQ must also be a fixed line, harmonically conjugated of OP with respect to d_1, d_2. This latter line is the required locus.

5.3.7. For the considered projective frame, we have $A[1:0:0]$, $B[0:1:0]$, $C[0:0:1]$, $M[1:1:1]$. Now, let A', B', C' be the midpoints of BC, CA, AB, respectively. Using Proposition 5.3.2, we get easily $A'[0:1:1]$, $B'[1:0:1]$, $C'[1:1:0]$. $B'C' \| BC$ means that $BC \cap B'C'$ is an improper point $M_\infty[m_1 : m_2 : m_0]$, where $m_1 = 0$, and where the coordinates are a linear combination of those of B', C' i.e.,

$$0 = \alpha + \beta, \ m_2 = \beta, \ m_1 = \alpha = -\beta.$$

It follows that $M_\infty[0:1:-1]$. Similarly we have $N_\infty := CA \cap C'A'[1:0:-1]$, and $P_\infty := AB \cap A'B'[1:-1:0]$. The improper line d_∞ of the plane is $M_\infty N_\infty$, for instance, and it has the parametric equations

$$x_1 = \mu, \ \ x_2 = \lambda, \ \ x_3 = -\lambda - \mu.$$

The elimination of the parameters provides the required equation of d_∞: $x_1 + x_2 + x_3 = 0$.

5.3.8. We will use projective coordinates such that $A[1:0:0]$, $B[0:1:0]$, $C[0:0:1]$, $D[1:1:1]$. Then, we have the equations (AB) $x_0 = 0$; (BC) $x_1 = 0$; (CA) $x_2 = 0$; (BD) $x_1 = x_0$; (CD) $x_1 = x_2$; (DA) $x_2 = x_0$, and $U[1:0:1]$. Now, we may see d as the line UX where $X[p:q:0]$, and (p,q) are homogeneous parameters. Then, every point of d is a linear combination of U and X. For instance, the general point $E = \lambda U + \mu X$, with homogeneous coefficients $[\lambda : \mu]$. We find the following values of these coefficients for the points which are interesting to us: $U[1:0]$, $X[0:1]$, $S[q:1]$, $Z[-p:1]$, $Y[q-p:1]$. Furthermore:

$$(U,S;E,X) = \frac{\mu q}{\mu q - \lambda}, \ (U,S;E,Z) = \frac{\mu(p+q)}{\mu q - \lambda}, \ (U,S;E,Y) = \frac{\mu p}{\mu q - \lambda},$$

and this proves the required equality.

5.3.9. Take A, B, C, D to be the fundamental points A_1, A_2, A_3, A_4 of the projective frame of space, and represent d by a pair of nonproportional equations (5.2.3)

$$f_1 := a_1x_1 + a_2x_2 + a_3x_3 + a_0x_0 = 0, \ f_2 := b_1x_1 + b_2x_2 + b_3x_3 + b_0x_0 = 0.$$

The solutions of $f_1 = 0$, $f_2 = 0$, $x_0 = 0$ are defined up to a factor, and we get

$$D'[(a_2b_3 - a_3b_2) : (a_3b_1 - a_1b_3) : (a_1b_2 - a_2b_1) : 0].$$

Then, similarly

$$A'[0 : (a_3b_0 - a_0b_3) : (a_0b_2 - a_2b_0) : (a_2b_3 - a_3b_2)],$$

$$B'[(a_3b_0 - a_0b_3) : 0 : (a_0b_1 - a_1b_0) : (a_1b_3 - a_3b_1)],$$

$$C'[(a_2b_0 - a_0b_2) : (a_0b_1 - a_1b_0) : 0 : (a_1b_2 - a_2b_1)].$$

If we project these points from CD onto AB, we get an equal cross ratio, whence

$$(A', B'; C', D') = \frac{(a_2b_0 - a_0b_2)(a_3b_1 - a_1b_3)}{(a_0b_1 - a_1b_0)(a_2b_3 - a_3b_2)}.$$

On the other hand, the plane $\alpha = (d, A)$ has the equation $\lambda f_1 + \mu f_2 = 0$ satisfied by $(1, 0, 0, 0)$, whence $[\lambda : \mu] = [-b_1 : a_1]$. Similarly, for $\beta = (d, B)$ we have $[\lambda : \mu] = [-b_2 : a_2]$, for $\gamma = (d, C)$ we have $[\lambda : \mu] = [-b_3 : a_3]$, and for $\delta = (d, D)$ we have $[\lambda : \mu] = [-b_0 : a_0]$. These values yield

$$(\alpha, \beta; \gamma, \delta) = \frac{\begin{vmatrix} -b_1 & a_1 \\ -b_3 & a_3 \end{vmatrix}}{\begin{vmatrix} -b_2 & a_2 \\ -b_3 & a_3 \end{vmatrix}} : \frac{\begin{vmatrix} -b_1 & a_1 \\ -b_0 & a_0 \end{vmatrix}}{\begin{vmatrix} -b_2 & a_2 \\ -b_0 & a_0 \end{vmatrix}} = (A', B'; C', D').$$

5.4.1. We have $A_1[1 : 0 : 0]$, $A_2[0 : 1 : 0]$, $A_0[0 : 0 : 1]$, and these coordinates satisfy the equation of Γ. The tangent of Γ at A_2 is $x_1 + x_0 = 0$, and the tangent of Γ at A_0 is $x_1 + x_2 = 0$. These two lines intersect at $P[1 : -1 : -1]$. The equation of the line AP as provided by (5.4.1) is $x_2 - x_0 = 0$, and this equation is satisfied by $[1 : 1 : 1]$. Finally, Γ has real points (e.g., A_1), and $\Delta = 1/4 \neq 0$ hence, Γ is a nondegenerate, real conic. We can also find its canonical equation. By the change

$$x_1 = y_1 + y_2, \quad x_2 = y_1 - y_2, \quad x_0 = y_0,$$

the equation of Γ becomes $y_1^2+2y_1y_0-y_2^2=0$ i.e., $(y_1+y_0)^2-y_0^2-y_2^2=0$. Then $z_1=y_0$, $z_2=y_2$, $z_0=y_1+y_0$, transforms the equation into $z_1^2+z_2^2-z_0^2=0$.

5.4.2. Choose projective coordinates with $A[1:0:0]$, $B[0:1:0]$, $C[0:0:1]$. Then Γ has the equation

$$2a_{12}x_1x_2+2a_{10}x_1x_0+2a_{20}x_2x_0=0.$$

The tangent at A is $a_{12}x_2+a_{10}x_0=0$, and it cuts (BC) $x_1=0$ at $P[0:a_{10}:-a_{12}]$. The tangent at B is $a_{12}x_1+a_{20}x_0=0$, and it cuts (AC) $x_2=0$ at $Q[a_{20}:0:-a_{12}]$. The tangent at C is $a_{20}x_2+a_{10}x_1=0$, and it cuts (AB) $x_0=0$ at $R[a_{20}:-a_{10}:0]$. And, the determinant of the coordinates of P,Q,R is zero.

5.4.3. If d is any line through $M(x_i^0)$, we may assume that the second point of d, used to get the intersection equation (5.4.5) is the harmonically conjugated point $N(x_i^1)$ of our problem, precisely. If we do so, and if we denote by $\mu=\lambda_1/\lambda_0$ a corresponding nonhomogeneous coordinate on d (λ_0,λ_1 are the parameters of (5.4.5)), we have $M(0)$, $N(\infty)$, and $P_1(\mu_1)$, $P_2(\mu_2)$ are the intersection points of d and Γ. The condition $(P_1,P_2;M,N)=-1$ means just $\mu_1+\mu_2=0$. Hence, the relations of the solutions and the coefficients of the quadratic equation (5.4.5) show that the coordinates x_i^1 of N must satisfy $f(x_i^0;x_i^1)=0$. This is exactly the required result.

5.4.4. The three collinear points have coordinates (x_i^1), (x_i^2), (x_i^3) $(i=1,2,0)$, where $x_i^3=\lambda x_i^1+\mu x_i^2$ for some $\lambda,\mu\in\mathbf{R}$. Accordingly, polarization of the equation $f(x_i)=0$ of the conic yields for the polar lines

$$f(x_i^3;x_i)=\lambda f(x_i^1;x_i)+\mu f(x_i^2;x_i)=0.$$

Since this is the equation of a pencil of lines, we are done.

5.4.5. Get the corresponding canonical equations by the Gauss method of square building.

5.4.6. We assume that Γ is real, and choose the projective frame such that the conditions of Exercise 5.4.3 hold, while $A_1[1:0:0]$, $A_2[0:1:0]$ are the given points. Then, the equation of Γ is (5.4.8) i.e., $x_1x_2-x_0^2=0$, and $\forall M\in\Gamma$ $(M\neq A_1,A_2)$, we have $M[tx_0:(1/t)x_0:x_0]$ $(x_0\neq 0)$.

Now, we look at the pencils of lines of centers A_1, A_2 as models of the projective line, and, between them, we establish the correspondence $A_1M \longmapsto A_2M$, where $M \in \Gamma$. The equations of these lines are

$$(A_1M) \quad tx_2 - x_0 = 0, \qquad (A_2M) \quad tx_0 - x_1 = 0.$$

Hence, if the lines $x_0 = 0$ (i.e., A_1A_2) and $x_1 = 0$ (i.e., the tangent of Γ at A_2) are taken as the basic lines of the first pencil, and $x_1 = 0$ (i.e., the tangent of Γ at A_1) and $x_0 = 0$ (i.e., A_2A_1) are taken as the basic corresponding lines of the second pencil, the established correspondence is just the equality of the interior, projective coordinates $[t : -1]$, in the two pencils. Therefore, the correspondence is a projective transformation, and it preserves the cross ratio. This is exactly the required result.

The result also holds for complex conics by the same computation with complex numbers. The result holds for a degenerate conic i.e., a pair of lines (d_1, d_2), if A_1, A_2 are assumed to belong both to the same line d_1 or d_2; then, either the cross ratios are not well-defined or we are in the case where the other points belong to the other straight line, and the result follows by invariance of the cross ratio by intersections and projections.

For the converse, let us look again at the pencils of centers A_1, A_2, where the latter are not the fundamental points of the frame, necessarily. In these pencils, we may use internal, projective frames with fundamental "points" A_1A_3, A_1A_4 and A_2A_3, A_2A_4, respectively, and with unit "points" A_1A_5, A_2A_5. This means that, in the plane, we have the following equations of these lines:

$$(A_1A_3) \quad d_1 = 0, \quad (A_1A_4) \quad d_2 = 0, \quad (A_1A_5) \quad d_1 + d_2 = 0,$$

$$(A_2A_3) \quad d'_1 = 0, \quad (A_2A_4) \quad d'_2 = 0, \quad (A_2A_5) \quad d'_1 + d'_2 = 0,$$

where d_i, d'_i are homogeneous, linear expressions of the form $\alpha_1x_1 + \alpha_2x_2 + \alpha_0x_0$. Now, the given cross ratio equality means that if A_1A_6 is $d_1 + \lambda d_2 = 0$, then A_2A_6 must be $d'_1 + \lambda d'_2 = 0$, with the same value of λ. Therefore, the locus of A_6 has the equation

$$d_2d'_1 - d_1d'_2 = 0,$$

which is clearly the equation of a conic Γ. The latter contains $A_1 = d_1 \cap d_2, A_2 = d'_1 \cap d'_2$. Γ is nondegenerate because of the condition imposed to the points $A_1, \ldots, A_5$. Without this condition, Γ might degenerate into two lines, one of which is A_1A_2.

5.4.7. As in the previous Problem 5.4.6, we put the equation of Γ under the form $x_1x_2 - x_0^2 = 0$, and $A_1[1:0:0]$, $A_2[0:1:0]$, $A_3[t_1 : 1/t_1 : 1]$, $A_4[t_2 : 1/t_2 : 1]$, $A_5[t_3 : 1/t_3 : 1]$, $A_6[t_4 : 1/t_4 : 1]$. Using formula (5.4.1) we get for the equation of A_4A_5:

$$(*) \qquad x_1 + t_2t_3x_2 - (t_2 + t_3)x_0 = 0,$$

and $P_1 = A_1A_2 \cap A_4A_5$ is $P_1[t_2t_3 : -1 : 0]$. (A_1A_2 has the equation $x_0 = 0$.)

Then, A_2A_3 has the equation $x_1 - t_1x_0 = 0$, and A_5A_6 has an equation similar to (*). This yields

$$P_2 = A_2A_3 \cap A_5A_6 \ \ [t_1t_3t_4 : (t_3 + t_4 - t_1) : t_3t_4].$$

Finally, A_3A_4 has an equation of the form (*), and A_6A_1 is $t_4x_2 - x_0 = 0$, whence

$$P_3 = A_3A_4 \cap A_6A_1 \ \ [-t_1t_2 + t_4t_1 + t_4t_2 : 1 : t_4].$$

Now, a straightforward computation shows that P_1, P_2, P_3 satisfy (5.4.1), and are collinear points.

If Γ is degenerated, the corresponding result is the Pappus' theorem of Problem 5.3.5.

5.4.8. i) A computation shows that the determinant of the coordinates of A, B, C, D is not zero. Hence, the points are not coplanar, and define a tetrahedron. The equation of the plane ABC is

$$\begin{vmatrix} x_1 & x_2 & x_3 & x_0 \\ 1 & -1 & 0 & 0 \\ 0 & 1 & -1 & 0 \\ 0 & 0 & 1 & -1 \end{vmatrix} = 0$$

i.e., $x_1+x_2+x_3+x_0 = 0$. Similarly, the plane ACD is $x_1+x_2-x_3-x_0 = 0$, and so on, in the same way we obtain the planes of the other faces, and the edge lines are the intersection of two such planes.

ii) We have $B \in \Gamma$, $D \in \Gamma$. The tangent planes of Γ at B, D are

$$x_2 + x_3 = 0, \qquad x_1 - x_0 = 0,$$

respectively.

iii) The vertices which do not belong to Γ are A and C. Following i), the planes through A, C form the pencil

$$\lambda(x_1 + x_2) + \mu(x_3 + x_0) = 0.$$

If the pole of this plane is (y_1, y_2, y_3, y_0), the plane must also be $y_1x_1 + y_2x_2 - y_3x_3 - y_0x_0 = 0$, and we must have

$$\frac{y_1}{\lambda} = \frac{y_2}{\lambda} = -\frac{y_3}{\mu} = -\frac{y_0}{\mu}.$$

Hence, $y_1 = y_2$, $y_3 = y_0$, and the required locus is the line of points which satisfy these two conditions.

5.4.9. The required point is the improper point of the image of the line $x_0 = 0$ by the transformation. The image of $x_0 = 0$ is obtained by inserting $x_0 = 0$ in the given equations and, then, eliminating x_1, x_2. The result is the line $8x'_1 + 6x'_2 - 3x'_0 = 0$, and its direction is $\bar{v}(-6, 8)$. Hence, the required improper point is $[-3 : 4 : 0]$, and it is the image of $[1 : 2 : 0]$.

5.4.10. The transformation has equations of the form indicated by formula (5.4.10):

$$(*) \qquad \rho x'_i = \sum_{j=0}^{2} a_{ij}x_j,$$

and we write them for each pair of corresponding points, each time with a corresponding value of the homogeneity factor ρ:

$$\rho_1 \cdot 0 = a_{11} \cdot 1 + a_{12} \cdot 0 + a_{10} \cdot 0 = a_{11},$$
$$\rho_1 \cdot 1 = a_{21} \cdot 1 + a_{22} \cdot 0 + a_{20} \cdot 0 = a_{21},$$
$$\rho_1 \cdot 0 = a_{01} \cdot 1 + a_{02} \cdot 0 + a_{00} \cdot 0 = a_{01},$$

i.e., $a_{11} = 0$, $a_{11} = \rho_1$, $a_{01} = 0$. Similarly, $a_{12} = 0$, $a_{22} = 0$, $a_{02} = \rho_2$, $a_{10} = \rho_3$, $a_{20} = 0$, $a_{00} = 0$.

Until now, we computed the coefficients a_{ij} by means of ρ_1, ρ_2, ρ_3. The last pair of corresponding points will determine ρ_1, ρ_2, ρ_3 up to a single homogeneity factor, say λ:

$$\lambda \cdot 1 = \rho_3 \cdot 1, \quad \lambda \cdot 2 = \rho_1 \cdot 1, \quad \lambda \cdot 4 = \rho_2 \cdot 1.$$

Hence, if we replace ρ by $\rho\lambda$ in (*), the equations of the transformation are

$$(**) \qquad \rho x'_1 = x_0, \quad \rho x'_2 = 2x_1, \quad \rho x'_0 = 4x_2.$$

Notice that this method is general but, if the given points have many nonzero coordinates, the computations will be longer and involve some real solving of linear equations. See an example in Problem 3 of the Appendix on the use of the Maple computer program.

As for the fixed points, in our case the characteristic equation (5.4.12) is $\rho^3 - 8 = 0$, and it has only one real solution $\rho = 2$. And, for $\rho = 2$, the characteristic system (5.4.11) of (**) yields the fixed point $[1 : 1 : 2]$.

5.4.11. The characteristic equation (5.4.12) of the given transformation is $\rho(\rho - 4)(\rho^2 - 2) = 0$, and it has the solutions $\rho_{1,2} = \pm 2$, $\rho_{3,4} = \pm i\sqrt{2}$. Only $\rho_{1,2}$ give us real fixed points, which are obtained by solving the corresponding, real, characteristic system (5.4.11). The real fixed points are $[\pm 1 : \pm 1 : 2 : 1]$.

5.4.12. We must look again for the fixed points of the given transformation. The characteristic equation is $\rho^3 - 3\rho + 2 = 0$. It has the solutions $\rho_1 = \rho_2 = -1$, $\rho_3 = 2$. For $\rho = -1$, the characteristic system reduces to only one independent equation $x_1 + x_2 + x_0 = 0$. This line consists of fixed points, and it will be the axis. The center is the fixed point of $\rho = 2$, and it is $[1 : 1 : 1]$.

5.4.13. Since O and K are fixed points, the line OK is fixed, and the image M' of $M \in OK$ also belongs to the line OK. Assume that $(O, K; M, M') = a$. Since a projective transformation preserves the cross ratios, and since φ is involutive (i.e., also $M = \varphi(M')$), we must have

$$a = (O, K; M, M') = (O, K; M', M) = 1/a.$$

Therefore $a = \pm 1$. But, since the four points are different $a \neq 1$, and we must have $a = -1$.

5.4.14. The same proof as for Problem 5.4.13 holds.

Bibliography

[1] Aleksandrov, P. S., Lectures on Analytical Geometry (in Russian). Nauka, Moscow, 1968.

[2] Berger, M., Geometry I, II. Universitext, Springer Verlag, Berlin, 1987.

[3] Berger, M., Pansu, P., Saint-Raymond, X., Problems in Geometry. Springer Verlag, Berlin, 1984.

[4] Char, B. W., Geddes, K. O., Gonnet, G. H., Leong, B. L., Monagan, M. B., Watt S. M., First Leaves, A Tutorial Introduction to Maple V. Springer-Verlag, New York, Berlin, 1993.

[5] Char, B. W., Geddes, K. O., Gonnet, G. H., Leong, B. L., Monagan, M. B., Watt S. M., Maple V Language Reference Manual. Springer-Verlag, New York, Berlin, 1992.

[6] Char, B. W., Geddes, K. O., Gonnet, G. H., Leong, B. L., Monagan, M. B., Watt S. M., Maple V Library Reference Manual. Springer-Verlag, New York, Berlin, 1992.

[7] Gheorghiev, Gh., Miron, R., Papuc, D., Analytical and Differential Geometry I (in Rumanian). Editura Didactică şi Pedagogică, Bucharest, 1968.

[8] Gurevich, G. B., Projective Geometry (in Russian). Gos. Izd. Fiz. Mat. Lit., Moscow, 1960.

[9] Horadam, A. F., A guide to undergraduate Projective Geometry. Pergammon, Rushcutters Bay, Australia, 1970.

[10] Mayer, O., Projective Geometry (in Rumanian). Editura Academiei, Bucharest, 1970.

[11] Penna, M., and Patterson, R. R., Projective Geometry and its Applications to Computer Graphic. Englewood Cliffs, N. J.: Prentice Hall, 1986.

[12] Postnikov, M., Lectures on Geometry I. Analytical Geometry (in French). Mir, Moscow, 1981.

[13] Salmon, G., A Treatise on the Analytical Geometry of Three Dimensions. Hodges, Dublin, 1874.

[14] Vaisman, I., Foundations of three-dimensional Euclidean Geometry. M. Dekker, Inc., New York, 1980.

[15] Vrănceanu, Gh., Analytical and Projective Geometry (in Rumanian). Editura Tehnică, Bucharest, 1954.

[16] Zuberbiller, O. N., Problems and exercices of Analytical Geometry (in Russian). Gos. Izd. Fiz. Mat. Lit., Moscow, 1958.

Index